FLYING
SQUADRONS
OF THE
AUSTRALIAN DEFENCE FORCE

STEVE EATHER

DEDICATION

This book is dedicated to the memory of
Warrant Officer Leonard Victor Waters
1924 – 1993

The RAAF's first Aboriginal pilot

CONTENTS

PREFACE

The purpose of this book is to record, in a readable fashion, a brief history of the Australian Defence Forces' major operational flying units – past and present.

Many small units equally deserving of consideration have not been included for space considerations. Of these, the histories of the RAAF units will be covered in a substantial series of publications to be released by the RAAF to mark that service's 75th Anniversary in 1996.

Researching this book has proved a long and difficult task. Not withstanding the RAAF's excellent collection of unit history records many discrepancies have been found and some points have proved well nigh impossible to verify. Obtaining information on the Navy and Army units has proved more difficult – in the later case extremely so. This has made the task of verifying information exceedingly hard. Notwithstanding these problems, any errors made are the sole fault of the author.

As with any book, many people provide assistance – in this case assistance so vital that without the support of the following people, this book would never have been finished: Mr R K Piper and Mr S Locke formerly of the RAAF Historical Section and latterly Mr D Wilson, Ms M Angel and Ms J Beck. Lieutenant Colonel R R Harding, Mr L Avery, Lieutenant Colonel D J Dowling, Lieutenant P MacKenzie, Lieutenant Commander R Geale, Curator of the Australian Naval Aviation Museum, Mr B Reardon, Curator of the Museum of Australian Army Flying, Warrant Officer D Gardner, Curator RAAF Museum, Ms M Walsh Librarian RAAF Museum.

Dennis Newton, a well known aviation historian, deserves special thanks for his thorough editing of the manuscript and pertinent comments. Dennis' assistance has been of immeasurable help as has that of John Bennett, Jim Thorn, Maria Davey and Gerard Frawley who turned my basic manuscript into a book. In conclusion, I must thank my wife Sue and children Michael, Andrew and Lauren, for their patience and support in what has proved a long and arduous task. Without their support I could not have gone on.

Steve Eather
June 1995

Published by Aerospace Publications Pty Ltd (ACN: 001 570 458), PO Box 3105, Weston Creek, ACT 2611, publishers of monthly *Australian Aviation* magazine.
Production Manager: Maria Davey

ISBN 1 875671 15 3

THE
AUSTRALIAN
FLYING CORPS

In January 1911, less than eight years after the Wright Brothers' first flight, the Australian Army began to ponder the usefulness of aircraft in the defence of Australia.

The resulting discussions progressed to the point where two experienced civilian aviators, Henry Petre and Eric Harrison, and two mechanics were recruited in the United Kingdom to form a flying school in Australia. The foundations of the Australian Flying Corps – Australia's first "Air Force" – were thus laid.

Henry Petre arrived in Australia in January 1913 and, after rejecting a proposed site at Duntroon, selected Point Cook on the shore of Port Phillip Bay near Melbourne as the site of Australia's first military flying school. Meanwhile Eric Harrison and the mechanics remained in the United Kingdom where Harrison ordered five aircraft for the proposed flying school; two BE2a biplanes, two Deperdussin monoplanes and a Bristol Boxkite.

The aircraft and associated support equipment – such as it was in those days – arrived at Point Cook in January 1914 and the Central Flying School, initially operating under canvas, came into existence. The first flight was made on March 1 and the School commenced pilot training with four students on August 18. The students all graduated but as there was no role for them in the embryonic air service, after being awarded their "Wings" they were returned to their original units.

The commencement of World War I bought with it a major increase in the strength and activities of the Australian Flying Corps. The small Corps undertook its first offensive operation when Lieutenant Harrison, one of the recently graduated pilots, Lieutenant G Merz, several mechanics and two aircraft joined an Australian expeditionary force which seized German New Guinea in late 1914. The Germans, however, offered only token resistance before surrendering and the aircraft were not used.

Next, on April 20 1915, what became known as the "Mesopotamian Half Flight" sailed for active service in the Tigris Valley where, equipped with obsolete British aircraft, it met with disaster and many of its personnel were either captured or killed.

Later in 1915, the British Government suggested that the various Dominions might wish to form their own complete flying squadrons for active service under British control. Only Australia acted upon this suggestion – the other Dominions being content to allow their personnel to serve in British squadrons (an arrangement in which Australia also participated).

Subsequently in January 1916, 1 Squadron formed at Point Cook and proceeded to the Middle East where it remained on active service until the end of the war. 1 Squadron was successively followed by 3, 2 and 4 Squadrons, all of which saw distinguished service on the Western Front. These four operational squadrons were backed by a United Kingdom based training wing comprising 5, 6, 7 and 8 Squadrons.

Having gained a distinguished record, all eight squadrons were disbanded in 1919 and their personnel returned to Australia. Through the Australian Flying Corps, however, were laid the traditions of achievement shortly to be taken up by the Royal Australian Air Force.

1 SQUADRON

The first complete unit of the Australian Flying Corps, 1 Squadron formed at Point Cook in January 1916 and, under the command of Lieutenant Colonel E H Reynolds, arrived in Egypt on April 14.

However, due to the lack of experience among its personnel, the aircrews were despatched to England for further flying training while the ground staff received "on the job" training with various Royal Flying Corps squadrons in the desert.

By June this training had been completed and 1 Squadron – which on September 12 was retitled 67 (Australian) Squadron, Royal Flying Corps – was ready for action. Officially based at Heliopolis, detached flights operated throughout the Canal Zone flying Bristol Scout, BE2a, BE2c, BE2e, BE12 and Martinsyde aircraft.

All these machines were obsolete, limited in performance and no match for the German aircraft opposing them. Notwithstanding these disadvantages, on April 12 reconnaissance and bombing missions commenced and were maintained despite many forced landings caused by engine failure.

The German and Turkish ground forces occasionally added to the Australians' discomfort and an example of this occurred on March 20 1917 when Lieutenant D W Rutherford's BE2c was forced down behind enemy lines after being hit by anti-aircraft fire. Despite himself being wounded by shell fragments, Lieutenant F

McNamara landed his Martinsyde to rescue the downed pilot.

Unfortunately, after collecting his colleague, McNamara's damaged aircraft crashed while taking off. In danger of being captured by approaching Turkish cavalry, the two aviators ran towards Rutherford's damaged aircraft which McNamara managed to start. The wounded Australian then took off under fire from the Turks and, despite severe blood loss, flew his fellow pilot back to base. For this selfless act, McNamara was later awarded the Victoria Cross – the only such award made to an Australian airman during World War I.

1 Squadron (still as 67 Squadron) finally concentrated at Mustabig, the first of a succession of bases, in December and continued operations in obsolete aircraft until October 1917 when modern RE8s were received. Before Christmas these aircraft, in turn, were supplanted by F2B Bristol Fighters – an aircraft of superlative performance. With these fighters 1 Squadron (having reverted from 67 to 1 Squadron in Jan 1918) played a pivotal role in wresting air superiority from the Germans in the Middle East. Lieutenant R A Austin claimed the squadron's first official kill, an Albatros scout on January 3 1918 – the first of 29 enemy aircraft to fall to the guns of the Australians.

Another vital task undertaken in this period was that of long range photo reconnaissance, which involved deep penetrations into enemy airspace. The unit consistently bought back high quality photographs from these operations and, such was the fear inspired by the Bristol Fighters, the Australians were rarely molested whilst so engaged.

Several notable individuals served in 1 Squadron. These included Lieutenant L J Wackett – later a founding member of the Australian aircraft industry, Captain R Smith – a brilliant aviator who, in 1919, would win the England-to-Australia air race, Lieutenant H Fysh one of the founders of the Queensland and Northern Territory Air Service – QANTAS – and Major Richard Williams – later Air Marshal Sir Richard Williams the RAAF's first Chief of Air Staff.

In September 1918, a large Handley Page 0/100 bomber (the only example of such an aircraft in the Middle East and the only twin engined aircraft operated by the AFC) flown by Captain Smith bombed and destroyed the El Afale telephone exchange and railway station during the opening phase of what would be the last major offensive against Turkey. As a direct result of this attack Turkish inland forces remained in ignorance of the allied offensive launched against

One of 1 Squadron's earlier types, the Martinsyde was not a particularly effective aircraft in any role allocated to it. (RAAF Museum)

their coastal armies and, consequently, were destroyed piecemeal. This aircraft, along with the Bristol Fighters, was also used to support the irregular Arab forces commanded by Captain T L Lawrence, the legendary "Lawrence of Arabia".

1 Squadron's Bristol fighters joined with British aircraft to bomb and strafe Turkish forces as they retreated towards Damascus, playing no small part in the enemy's further demoralisation. On September 22, a patrolling Bristol Fighter located a very large number of Turkish troops withdrawing up the Waddy Fara. As a result of this sighting report, relays of Australian and British aircraft attacked the Turks in the steep valleys where they had no chance of escape and few places to hide.

Later known as the Battle of Armageddon, three Turkish armies were so badly cut up by the bombing and strafing that they ceased to exist as an effective fighting force. Finally, on October 31 1918, Turkey surrendered and after moving back to Kantara, 1 Squadron disbanded on May 5 1919.

Formed: *January 1916*
Aircraft: *BE2, BE12, DH.6, Martinsyde G.100, Martinsyde G.102, Bristol Scout, Nieuport 17, RE8, Bristol Fighter, Handley Page 0/400*
Locations: *Point Cook: January – March 1916*
Heliopolis: April – December 1916
Mustabig: December 1916 – January 1917
Kilo 143: January 1917 – March 1917
Rafa: March – June 1917
Deir-el-Belah: June – September 1917
Wehi Sheikh Nuran: Sept – Dec 1917
Julis: December 1917 – February 1918
El Mejdel: February – April 1918
Ramleh: April – October 1918
Haifa: October – November 1918
Ramleh: November 1918 – February 1919
Kantara: February – March 1919
Disbanded: *May 5 1919*

2 SQUADRON

2 Squadron formed at Kantara, Egypt, on September 20 1916. Commanded by Major O Watt the new fighter squadron arrived at Harlaxton in the United Kingdom in January 1917 by which time it had been retitled 68 (Australian) Squadron, Royal Flying Corps by the British.

Having completed its operational training, the unit's aircraft flew to France on September 21 1917. From the first it seems that 68 Squadron was destined to be more than "just another squadron" as the safe arrival of all its aircraft at St Omer from the then still adventurous cross channel flight in one day was a first in itself for the "British" air service.

Initially based at Baizieux, 68 Squadron commenced combat flying on October 21. The DH.5s with which the unit was equipped, were not especially good aircraft due to their poor high altitude performance and, consequently, they were mainly operated in the ground attack role.

Flying at low level in the face of heavy ground fire, the Australian pilots wreaked considerable damage on the German frontline troops but themselves suffered a steady stream of casualties while carrying out this most dangerous work. During this period almost every aircraft was repeatedly damaged and over a third of them destroyed. Losses in pilots were correspondingly heavy although a surprising number survived as POWs.

Lieutenant F Huxley claimed 2 Squadron's (then still 68 Sqdn) – and indeed the AFC's – first aerial victory on November 22 when he shot down an Albatros scout in the course of a ground strafing mission. Several German aircraft were destroyed before 2 Squadron was withdrawn from operations to re-equip with SE.5As in December 1917.

A slightly damaged 2 Squadron SE5a. Note the distinctive boomerang fuselage marking. (RAAF Museum)

By mid January 1918 when the unit reverted to 2 Squadron, it was again operational and, flying from Savy, commenced high altitude patrols over enemy lines. The return of 2 Squadron to combat flying also coincided with the unit regaining its correct identity as a squadron of the Australian Flying Corps.

The SE.5As gained their first successes in February and the number of air combat engagements increased rapidly during March when the Germans opened their famous 1918 offensive. 2 Squadron was forced to withdraw from several airfields in the face of the enemy advance but operations continued unchecked and the Australians inflicted heavy losses on the opposing German squadrons. During one two day period in May, 2 Squadron pilots, for the loss of one SE.5A, shot down almost the numerical equivalent of a squadron of aircraft!

To help blunt the German offensive, 2 Squadron flew many low level strafing and bombing attacks against both the forward troops and those providing their logistical support further in the rear areas. These operations inflicted considerable loss on the enemy who, at times, were thoroughly disorganised by these aerial assaults. Return fire was heavy and resulted in the loss of several aircraft and damage to others. During this period it is reputed that Captain L Holden, one of the unit's most daring pilots, used up on average one aircraft per day!

By the end of June, 2 Squadron was based at Reclinghem where it worked in close co-operation with 4 Squadron conducting large scale fighter sweeps. On August 16 and 17, 2 Squadron participated in massed air attacks on Habourdin and Lomme airfields. Both targets were virtually destroyed and losses in aircraft

and equipment were heavy. So successful proved these air raids that further operations of this type were mounted prior to the close of hostilities. In the final weeks of the war, 2 Squadron was mainly occupied attacking the retreating German Army. Moving three times in October, the signing of the Armistice found 2 Squadron based at Pont-a-Marq. The unit relocated to Hellemmes in December, and then to Hurdcott in the UK in March, where it was disbanded on May 6 1919.

During the war, 2 Squadron spawned a total of 18 ace pilots, the highest scorer being Major R C Phillipps. This officer, despite having a partially paralysed leg (a result of earlier service with the infantry), destroyed 15 aircraft including four in one flight. Between them, 2 Squadron pilots shot down 94 enemy aircraft, sent another 73 down out of control and drove down a further 18.

Formed: *September 20 1916*
Aircraft: *DH.5, SE.5A*
Locations: *Kantara: September 1916 –*
Harlaxton: January – September 1917
Baizieux: September 1917 – January 1918
Savy: January – March 1918
La Bellevue: March 1918
Fouquerolles: March – June 1918
Liettres: June – October 1918
Serny: October 1918
Auchel: October 1918
Pont-a-Marq: October – December 1918
Hellemmes: December 1918 – February 1919
Hurdcott: March – May 1919
Disbanded: *May 6 1919*

3 SQUADRON

Formed at Point Cook on September 19 1916 under the command of Major D V J Blake, 3 Squadron became the first Australian Flying Corps unit to arrive in Europe when, in December, it moved to England for further training.

During this period it operated BE2e, Curtis Jenny and Armstrong Whitworth FK8 aircraft and was redesignated 69 (Australian) Squadron, Royal Flying Corps. 3 Squadron was to carry this designation until January 19 1918 when it regained the title assigned to it by the Australian Imperial Force.

Equipped with RE8 aircraft, 3 Squadron moved to Savy in France during September where its role was to conduct artillery shoots as well as bombing, strafing, tactical reconnaissance and photographic missions in support of British and Canadian troops. In November 1917, 3 Squadron moved to Baileul in Flanders where it became the corps reconnaissance squadron to the 1st ANZAC Corps at Messines and as such supported Australian troops for the remainder of the war.

On December 6, 3 Squadron achieved its first aerial victory when Captain W Anderson's observer, Lieutenant J Bell, shot down a German observation aircraft. Before the end of the war another 15 enemy aircraft would fall to the guns of the Australian unit while others were damaged or forced to land.

The most remarkable thing about this string of victories was that the slow RE8s actually sought out combat opportunities with enemy aircraft and frequently chased their more nimble adversaries in a bid to close with them. Notwithstanding these successes, some RE8s were bought down by German fighters or anti-aircraft fire. In one or two other incidents RE8s are believed to have been "shot down" after being hit in midair by artillery shells from batteries they were directing.

One of the most unusual incidents of World War I occurred on December 17 1917 when an RE8 crewed by Lieutenant J Sandy and Sergeant H Hughes was attacked by six Albatros scouts.

Sergeant Hughes wounded the pilot of one enemy aircraft which was forced to land behind Allied lines. With the arrival of another Australian RE8 the remaining Germans broke off the action. Noticing Sergeant Hughes waving to them, the other crews presumed everything was all right and returned to their own tasks. Nothing further from the truth could have been the case however, as in the last minutes of the engagement, Sergeant Hughes had been shot and mortally wounded by a bullet which penetrated through his chest before striking Lieutenant Sandy in the head, killing him instantly. Their inherently stable and only slightly damaged RE8 continued to fly in wide left hand circles until it finally ran out of fuel and landed relatively intact in a snow drift some 80km from the scene of the combat.

The bodies of the two Australians were recovered next day. The Albatros bought down by Sergeant Hughes in this combat was subsequently presented to Australia as a war trophy and is now on display at the Australian War Memorial, Canberra.

In March 1918, 3 Squadron was forced from its well established airfield at Baileul by German long range shelling and air attack. These attacks were preliminaries to the German's "March" offensive which was intended to end the war in Germany's favour. Initially relocating to Abeele, several other moves were forced upon the unit as German troops continued to break through the British defence lines. Despite the disruption caused by these moves, the RE8s continued to conduct effective operations in support of their Australian ground forces.

3 Squadron was involved in another unusual event on April 21 1918 when two of its aircraft on a photographic mission were attacked by four German fighters led by Baron Von Richtofen (the so called 'Red Baron' and the highest scoring ace of World War I). Despite the fact these German pilots were supposed to be the elite of the German air service, the Germans – including the Baron himself – were driven off by the lumbering Australian reconnaissance machines.

Looking for easier pickings the Baron made the mistake of descending to low level over the ANZAC Corps frontline and was shot down and killed by ground fire. The Baron's body was subsequently recovered by 3 Squadron and buried with full military honours. The remains of his crimson red triplane was held in custody by 3 Squadron until it could be handed over to British authorities.

On June 9 1918, 3 Squadron again figured in an unusual incident when an RE8 flown by Lieutenant R C Armstrong caught a Halberstadt aircraft flown by a young inexperienced crew. Not wishing to be shot down, the Germans allowed themselves to be shepherded to 3 Squadron's airfield under the watchful eyes (and guns) of the

Australian crew where they landed and were arrested. The intact aircraft was later presented to Australia as yet another war trophy.

3 Squadron continued to support the Australian Army for the remainder of 1918 and, to support the Allied counter offensive, refined the aerial delivery of ammunition to forward troops. The parachute dropping of ammunition proved most successful and resulted in these vital stores being delivered to forward positions much more quickly than had been the case when undertaken by ground troops (who often suffered heavy casualties in the process).

Just before war's end 3 Squadron began to re-equip with Bristol fighters. After the Armistice the unit which, with some justification, had been regarded as the best corps reconnaissance squadron on the Western Front, provided an aerial mail service for the Allied armies before disbanding in February 1919.

Formed: *September 19 1916*
Aircraft: *Curtis Jenny, Armstrong Whitworth FK8, BE2e, RE8, Bristol Fighter*
Locations: *Point Cook: Sept – October 1916*
South Carlton: Dec 1916 – Sept 1917
Savy: September – November 1917
Baileul: November 1917 – March 1918
Abeele: March – April 1918
Poulainville: April – May 1918
Villers-Bocage: May – September 1918
Proyart: September 1918
Bouvincourt: September – October 1918
Montigny: October 1918
Premont: October – December 1918
Charleroi: December 1918 – February 1919
Disbanded: *February 28 1919*

One of 3 Squadron's ever reliable RE8s on a French landing ground. (via RAAF Museum)

4 SQUADRON

4 Squadron, Australian Flying Corps, formed at Point Cook on October 25 1916, under the command of Captain A Lang.

Upon arriving at Yatesbury in the United Kingdom to complete its training, the unit was renamed 71 (Australian) Squadron, Royal Flying Corps, and it was not until January 19 1918 that its correct title was reallocated.

On being declared operational, the squadron moved to Bruay, France, on December 18 1917 and commenced combat operations with its Camels in mid January 1918. The novice fighter squadron was blooded on January 24 when Captain A H O'Hara-Wood shot down a DFW two seater. On February 3, more victims fell to the Australian Camels when three enemy aircraft were shot down. Lieutenant A H Cobby – an officer later to become the AFC's most successful fighter pilot – claimed his first kill in this engagement and the other successful pilots were Captain O'Hara-Wood and Lieutenant E F Pflaum.

From the beginning, 4 Squadron occasionally met the elite 'Richtofen's Flying Circus' and, in spite of their lack of experience, quickly gained the ascendancy over this German unit, repeatedly causing them loss. In its first clash with the 'Flying Circus' on March 16 five Camels engaged 16 Albatros scouts, one Australian pilot was shot down and spent the rest of the war as a POW. However, two – possibly three – chairs in Richtofen's mess were empty that night.

In March, the Germans launched their '1918 Offensive'. To help counter the initial German success, many hazardous ground attack missions were undertaken. 4 Squadron inflicted heavy losses on the German infantry in this period, however, many Camels were damaged, and some lost to ground fire. By the end of April, 4 Squadron had moved to Clairmaris after Bruay came under concentrated artillery fire. The intensive flying effort required in this period placed a great strain on all ranks – particularly the ground staff who had to repair, rearm and refuel aircraft as quickly as possible.

In mid May, 4 Squadron suffered a major setback when, after participating in a large scale attack against Armentiers in which one pilot was shot down, low fog covered the countryside. The Camels were forced to put down through the murky fog and another six were destroyed in crash landings.

Fokker DV.II fighters were encountered by 4 Squadron for the first time in mid 1918. These new aircraft had superior performance to the Camels, especially at high altitude. Such, however, was the skill and determination of the Australian pilots, that these new aircraft regularly fell to the guns of 4 Squadron. The Australian squadron could not have it all its own way, however, and on September 5 a patrol of five Camels was attacked by a very large number of enemy aircraft. In the unequal combat which followed, four Camels were shot down resulting in the deaths of three pilots – the fourth survived as a POW.

One method of countering the Fokker's superior performance was for the Camels to fly at low level in mixed formations with aircraft above

A 4 Squadron Sopwith Camel. (RAAF Museum)

them possessing better high altitude performance such as the SE.5A and Bristol Fighters. These large patrols, in which each aircraft type flew at its optimum height, allowed 4 Squadron to operate successfully against the German fighters. Some of these combined patrols were made in conjunction with the SE.5As of 2 Squadron and resulted in a very competitive spirit developing between both units.

In October 4 Squadron re-equipped with high performance Sopwith Snipes which enabled it to meet the new Fokker on equal terms for the remaining weeks of the war. During this period, there were many large scale air battles and a considerable number of enemy aircraft were bought down. On only one occasion did the Germans manage to inflict substantial losses on 4 Squadron while operating Snipes. This occurred on November 4 when five Snipes were shot down when the unit engaged a large German formation in what proved to be 4 Squadrons last major battle of the war.

After Germany's capitulation, 4 Squadron moved to Cologne as part of the Allied occupation force. This was not last for long however as, in February 1919, the unit disbanded and its personnel returned to Australia. During its brief war service, 4 Squadron had destroyed some 128 enemy aircraft – with many others driven down or damaged – and spawned a total of 11 aces, the highest scoring being Captain A H Cobby who, in addition to shooting down 29 aircraft, also destroyed 13 observation balloons.

Formed: *October 25 1916*
Aircraft: *Sopwith 1½ Strutter, BE2e, BE12, Camel, Snipe*
Locations: *Point Cook: Oct 1916 – Jan 1917*
Castle Bromwich: April – December 1917
Bruay: December 1917 – April 1918
Clairmarais: April – June 1918
Reclinghem: June – September 1918
Serny: September – October 1918
Auchel: October 1918
Seclin: October 1918
Grand Ennetieres: October – December 1918
Euskirchen: December 1918
Bickendorf: December 1918 – February 1919
Disbanded: *February 28 1919*

---

5 SQUADRON

5 Squadron formed at Shawbury, England on September 1 1917. The new unit – known initially as 29 (Australian) (Training) Squadron RFC – had for its first commander a most experienced airman in the person of Major H A Petre. It was this officer who had done much to establish military aviation in Australia and had later commanded the "Half Flight" during the ill fated 1915 Mesopotamian expedition.

Initially equipped with Maurice Farman Shorthorns, 5 Squadron late received DH.6, Avro 504 and Sopwith Pup trainers and operated as an initial training school passing its pupils on to other units for more advanced flying training after they had completed around ten hours tuition.

A DH.6 trainer of the type used by 5 Squadron in the United Kingdom. (via RAAF Museum)

On January 18 the unit was officially renumbered 5 (Training) Squadron, Australian Flying Corps and, with a redesigned training syllabus, commenced "all through" training courses later in the year. To facilitate the more advanced phases of this training, the unit took delivery of SE.5As and Camels. Mainly providing replacement personnel for 4 Squadron, 5 Squadron graduated an average of eight pilots each month and utilised the services of many combat experienced pilots as instructors.

One of the most admired instructors in 5 Squadron was Captain R King, who with a personal score of 16½ enemy aircraft destroyed, was ranked the Australian Flying Corps' third highest ace pilot.

5 Squadron moved to Minchinhampton on April 2 1918 where pilot training continued using the advanced Snipe. In May 1919, 5 Squadron disbanded at Minchinhampton.

Formed: *September 1 1917*
Aircraft: *Shorthorn, DH.6, Avro 504, Pup, SE.5A, Camel, Snipe*
Locations: *Shawbury: Sept 1917 – April 1919*
Minchinhampton: April 1918 – May 1919
Disbanded: *May 1919*

6 SQUADRON

6 Squadron formed at Yatesbury in the United Kingdom in September 1917. The new unit, under the command of Major A A J Brown, was known until January 1918 as 30 (Australian) (Training) Squadron, Royal Flying Corps.

6 Squadron undertook the training of fighter pilots utilising a variety of aircraft types including the Bristol Scout D, Sopwith 1½ Strutter, Sopwith Pup, Avro 504, DH.5, SE.5A and Sopwith Camel F1.

From Yatesbury, 6 Squadron moved to Tern Hill and hence to Minchinhampton in February 1918. Pilot training continued after the end of the war and the unit finally disbanded in March 1919.

Formed: September 1917
Aircraft: Bristol Scout, Pup, DH.5, Sopwith 1½ Strutter, Avro 504, SE.5A, Camel
Locations: Yatesbury: Sept 1917 –
Tern Hill:
Minchinhampton: February 1918 – March 1919
Disbanded: March 1919

(right) An unarmed SE.5A of an Australian training squadron. (RAAF Museum)

7 SQUADRON

Commanded by Captain H D E Ralfe, 7 Squadron formed at Yatesbury, England on October 24 1917. Initially equipped with DH.6, BE2e and RE8 aircraft, 7 Squadron did not commence its flying training role until moving to Leighterton in February 1918 and, until this time, was known as 32 (Australian) (Training) Squadron, RFC.

At its new base, 7 Squadron received Avro 504s and later F2B Bristol Fighters (which eventually replaced the DH.6 and RE8) and was tasked to supply replacement aircrews to 3 Squadron. As such, many of the unit's instructors were highly experienced airmen who had completed an operational tour of duty with 3 Squadron on the Western Front. Continuing to operate after the Armistice in September 1918, 7 Squadron disbanded on an unrecorded date in early 1919.

Formed: October 24 1917
Aircraft: BE2e, DH.6, Avro 504, RE8, Bristol Fighter
Locations: Yatesbury: October 1917 – Feb 1918
Leighterton: February 1918 – 1919
Disbanded: 1919

Starting up an Avro 504 for another pilot training flight. (RAAF Museum)

8 SQUADRON

8 Squadron formed at Yatesbury on October 25 1917. The new Australian training squadron, commanded by Major G A C Cowper was initially known as 33 (Australian) (Training) Squadron, RFC.

Still without aircraft, 8 Squadron moved to Cirencester and then to Leighterton in January 1918 and there received Avro 504s, Sopwith Pups, Sopwith Camels and later, Sopwith Snipes. With these aircraft it conducted pilot training in support of 4 Squadron. To carry out this important work, 8 Squadron utilised the services of a number of experienced fighter pilots – the most notable being Captain A H Cobby – the AFC's highest scoring fighter pilot.

The signing of the Armistice heralded the end of World War I but, like the other training units, 8 Squadron continued to operate for a time before disbanding at Leighterton on April 30 1919.

Formed: October 25 1917
Aircraft: Pup, Camel, Avro 504, Snipe
Locations: Yatesbury: October 1917 – Jan 1918
Cirencester: January 1918 – February 1918
Leighterton: February 1918 – April 1919
Disbanded: April 30 1919

With its distinctive Kangaroo markings, this 8 Squadron Avro 504 is seen at either Minchinhampton or Leighterton. (RAAF Museum)

THE
ROYAL
AUSTRALIAN
AIR FORCE

On March 31 1921 the Royal Australian Air Force came into existence. The initial strength of the new service, which, upon the date of its formation, was only the second independent air force in the world, comprised 21 officers and 130 airmen, with around 160 aircraft.

Personnel for the new service were mainly veterans of World War I who had either served in the Australian Flying Corps or one of the British air services. Commanding the RAAF was Wing Commander Richard Williams (later Air Marshal Sir Richard Williams), an officer with a distinguished war record and the first pilot to graduate at Point Cook in 1914.

In a period of severe budgetary constraints followed by a world depression, the RAAF grew very slowly and it was only in 1925 that the first two flying squadrons were formed. Initially, only the base at Point Cook was available but during the 1920s further bases were established at nearby Laverton and Richmond, north west of Sydney.

Not until the mid 1930s, with the growing threat of war in Europe, did the Government begin to allocate adequate funds for the purchase of substantial numbers of aircraft, the construction of airfields and the establishment of additional squadrons. On September 3 1939, the day that Australia declared war on Germany, the RAAF consisted of just 3489 personnel and 246 aircraft – almost all of them obsolete.

While still a small air force, the RAAF began a period of rapid expansion and, apart from training personnel for its own requirements, also provided many thousands of air and ground crew for service with the Royal Air Force under the provisions of the Empire Air Training Scheme. Australian airmen, serving in either RAAF or other Commonwealth squadrons, fought in every theatre of the war. From December 1941 RAAF squadrons were engaged against Japanese forces and, for the first time, were involved in combat operations mounted in the direct defence of Australia.

After the surrender of Japan in August 1945, the RAAF had become the fourth largest air force in the world and, at its peak, operated over 8000 aircraft. Total wartime enlistments were 216,900 men and women of whom 10,562 were killed and another 3192 injured. A total of 76 flying squadrons and a very large training and support organisation had been formed.

Despite commencing a demobilisation program of massive proportions (which in a short period would reduce the RAAF to just 8000) the RAAF was immediately committed to further overseas activities. From early 1946, a fighter wing comprising three Mustang squadrons and nearly 2000 personnel participated in the occupation of Japan.

One of these squadrons (77) was still in Japan when the Korean War commenced in June 1950. The commitment of this squadron resulted in Australia becoming the first United Nations country outside the United States to undertake active operations in Korea. A transport squadron was also committed to support United Nations forces in Korea while parallel to this, a bomber and transport squadron undertook active operations against communist terrorists in Malaya.

Earlier, in 1948/49 the aircrew equivalent of a transport squadron had been delivering supplies into besieged Berlin, during the Berlin Airlift. The 1950s also saw a further two fighter squadrons based in Malta on garrison duties for a period of two years.

By the beginning of the 1960s the RAAF was firmly established in the jet age and was operating a balanced force both in Australia and at Butterworth, Malaysia. At Butterworth two fighter squadrons, a bomber squadron and a transport support unit were located, giving Australia a forward defence posture in Asia.

In 1964, Caribou transport aircraft were committed to action in Vietnam and these were later followed by Iroquois helicopters and Canberra jet bombers, plus various Army Aviation aircraft. More recently, during the 1970s, '80s and '90s, the RAAF has participated in a number of UN peacekeeping operations, where the conduct of its personnel have continued to add lustre to the RAAF's already impressive reputation.

Today the RAAF operates a mixture of aircraft capable of providing the effective defence of Australia. Participation in regular international exercises continues to hone the service's fighting skills to a point where, although small, the RAAF remains an exceedingly effective force respected throughout the world.

Commanded by Flight Lieutenant A H Cobby – the highest scoring AFC pilot of World War I – 1 Squadron formed at Point Cook on July 1 1925. Equipped with SE.5A, DH.9 and DH.9A aircraft it operated as a composite squadron with each of its three flights training in a separate role – fighter, bomber and army co-operation. From the start, two thirds of the unit's personnel were to be Citizen Air Force members.

Later re-equipped with Wapitis (and subsequently Bulldogs and Demons), 1 Squadron, which moved to Laverton in June 1928, participated in a number of civil aid tasks and airshows during the 1920s and 1930s. These activities, which generated a tremendous amount of publicity for the fledgling RAAF, were in addition to its normal military training.

By the commencement of World War II, 1 Squadron had re-equipped with Avro Anson Mk.Is and, utilising these aircraft, conducted routine convoy escort and patrol missions prior to their being replaced by Lockheed Hudsons in May 1941. The unit then moved to Singapore where it was initially based at Sembawang on Singapore Island and later at Kota Baharu in the north of Malaya.

Reconnaissance operations and training continued up until December 6 when patrolling Hudsons located Japanese convoys approaching Malaya. Sightings were also made the next day but, on the orders of higher authority, no action was taken.

In the early hours of December 8 (local time), 1 Squadron personnel heard the unmistakable sound of naval gunfire being directed against the defences on Kota Baharu beach. The garrison commander informed 1 Squadron that a Japanese landing was in progress and the unit commenced immediate air strikes against the invasion force. By morning, two ships had been sunk and one damaged in addition to a number of landing barges. In return for these losses two Hudsons had been shot down and several others damaged.

Attacks against the enemy continued into the morning – despite air raids on Kota Baharu – and further damage was inflicted on the Japanese. Nonetheless, by the end of the first day, the Japanese were firmly established and only five serviceable Hudsons remained to resist them. Undoubtedly the Japanese landing had been successful and, as the Army garrison was overcome, Kota Baharu airfield came under small arms fire and had to be abandoned that afternoon.

1 Squadron withdrew its few remaining aircraft to Kuantan and thence to Sembawang where it was amalgamated with the remnants of 8 Squadron. Reconnaissance and bombing missions – sometimes made by night against targets in Malaya – continued until January 29 when, after taking over 8 Squadron's remaining Hudsons, the unit moved to Palembang on Sumatra.

On February 14, the Japanese invaded Sumatra. These forces were repeatedly bombed and strafed but, despite very damaging attacks, succeeded in gaining a foothold on the island. Two days later, 1 Squadron retreated to Palembang on Java. Attacks against the Japanese continued up until the first days of March when the unit's four surviving Hudsons were flown to Australia. 1 Squadron's remaining personnel, along with other Commonwealth, Dutch and American

An SE.5A about to start up with the aid of a "Huck Starter". (RAAF Museum)

An Australian built Lincoln heads towards its target during the Malayan Emergency.

troops were then forced to surrender to the Japanese. Of the 160 Australian airmen to fall into enemy hands less than half were to survive their years of brutal captivity.

1 Squadron reformed at Menangle on December 1 1943 and, equipped with Beauforts, moved to Gould in the Northern Territory. Reconnaissance and bomber operations, particularly over Timor, commenced on March 20 1944. Many of these missions were flown by night and losses – mainly from anti-aircraft fire – were not unduly heavy. Nonetheless there were exceptions and on May 8 1944 two out of three Beauforts despatched against Penfui failed to return. The Beauforts also, at times, participated in joint attacks with Mitchells of 18 Squadron.

Less exciting, but certainly no less important, were the many blue water patrols carried out in this period. Most of these missions were uneventful and there was little to report. Nonetheless, these "routine" flights bought with them their own special hazard – the danger of suffering engine failure while far out to sea.

In January 1945, 1 Squadron relocated to Kingaroy and converted to Australian built Mosquito FB.VI attack aircraft prior to moving to Moratai in May. This was followed by a further move to Labuan in June/July 1945. Due to a lack of suitable targets however, only a few Mosquito missions were flown prior to the war's end. 1 Squadron returned to Australia where it disbanded at Narromine on August 7 1946.

Equipped with Lincoln Mk.30 bombers, 1 Squadron again reformed on February 23 1948 and after two years at Amberley, moved to Tengah in July 1950 where it participated in operations against Communist Terrorists in Malaya. For the next eight years, the Lincolns pounded enemy hideouts, usually without any observed results.

As far as confirmed kills went, 1

Squadron's most successful mission of the Malayan Emergency – 'Operation Kingly Pile' – was flown on February 21 1956 when seven Lincolns pattern bombed a hideout north west of Kluang killing 18 terrorists.

Initially the Lincolns operated by day – their targets small nondescript patches of jungle which required pinpoint navigation to locate, much less attack. The jungle covered targets were often designated with the aid of an airborne forward air controller's smoke grenades, artillery smoke or by making timed runs from distinctive datum points – such as a hill or prominent bend in a river.

After almost every daylight attack the Lincolns descended to treetop height while their gunners thoroughly strafed the target area with 20mm cannon and .50 calibre machinegun fire. Some regarded the strafing as being more effective than the bombing itself but the pilots really did not care as these attacks gave them plenty of legal opportunities to engage in low level flying. More than one Lincoln returned to Tengah with tree branches lodged in its engine nacelles!

As the Malayan Emergency progressed, 1 Squadron began to dispatch single bombers on moonlit nights to harass the Communists around the clock. These missions were of much longer duration than daylight attacks as, often, only

Both 1 and 6 Squadrons operated Phantoms in the early 1970s. (RAAF)

A bombed up F-111 over southern New South Wales. (RAAF)

single bombs were dropped at irregular intervals. As the crews gained experience and confidence in night bombing techniques, formations of three and later five bombers were despatched on these nocturnal attacks. It is significant to note that 1 Squadron was the only unit engaged in the Malayan Emergency to operate by night.

With the Malayan Emergency all but concluded and having dropped no less than 85% of all the bombs dropped during the campaign, 1 Squadron returned to Australia in July 1958 and immediately converted to Canberra Mk.20 twinjet bombers. With these Australian built jets, 1 Squadron participated in numerous exercises, and overseas deployments – destinations including New Zealand, Cocos Islands and various Pacific localities – prior to converting to McDonnell Douglas F-4E Phantom fighter bombers in late 1970. The Phantoms – 1 Squadron's first supersonic aircraft – despite their superlative air-to-air combat capability were used almost exclusively in the strike role.

Although only operated for two years, the modern technology represented by the Phantoms was to provide very useful experience to both air and ground personnel awaiting the delivery of the long delayed F-111 strike aircraft. So popular were the Phantoms, in fact, that their crews were reluctant to part with them when they were returned to the United States during October 1972.

1 Squadron finally received its first General Dynamics F-111Cs in June 1973 and continues to operate these aircraft (along with 6 Squadron) as Australia's primary defence deterrent. Over the years, the F-111s have been modified and modernised and now utilise the Pave Tack electro optical and laser guided bombing system for pin point bombing attacks while the lethal Harpoon long range missile is carried when the aircraft are operating in the anti-shipping role.

The F-111Cs of 1 Squadron participate in regular defence exercises including many overseas in Malaysia, New Zealand, the Philippines and the United States. With an expected remaining service life until 2015, 1 Squadron will continue to operate these superlative attack aircraft well into the 21st Century.

Formed: July 1 1925
Squadron Codes: A, US, NA
Aircraft: SE.5A, DH.9, DH.9A, Wapiti, Bulldog, Demon, Anson, Hudson, Beaufort, Mosquito, Lincoln, Canberra, Phantom, F-111C
Locations: Point Cook: July 1925 – June 1928
Laverton: June 1928 – July 1940
Sembawang: July 1940 – August 1941
Kota Baharu: August – December 1941
Sembawang: Dec 1941 – January 1942
Palembang: February 1942
Semplak: February 1942
Kalidjati: February – March 1942
Andir: March 1942
Disbanded:
Menangle: December 1943 – March 1944
Gould: March 1944 – January 1945
Kingaroy: January – May 1945
Moratai: May – June 1945
Labuan: June – December 1945
Narromine: December 1945 – June 1946
Disbanded:
Amberley: February 1948 – July 1950
Tengah: July 1950 – July 1958
Amberley: July 1958 – present

------

2 Squadron formed at Laverton on May 3 1937. Initially equipped with Demon Is and commanded by Squadron Leader J H Summers, the unit later operated a mixed fleet of aircraft which, aside from the Demons, comprised Anson Mk.Is, Bulldog IIAs, and Wirraways.

By the time that war was declared in September 1939, however, 2 Squadron was solely operating Ansons. These aircraft conducted maritime patrol and convoy escort missions – often covering ships carrying Australian troops to the Middle East. Searches for German surface raiders, thought to be operating off Australia's east coast and especially in Bass Strait, were also a regular feature of 2 Squadron's work.

In June 1940, 2 Squadron began to re-equip with Hudsons. After receiving its full allocation of these aircraft, routine patrols and exercises continued until the first week of December 1941 when the unit deployed to Darwin in anticipation of Japan's entry in the war. Almost immediately detached flights of Hudsons moved forward to Penfaei and later Ambon where advanced operational bases were established.

Early on the morning of December 8, Flying Officer Law-Smith's crew took off on a reconnaissance mission. As it was more than likely that active operations against the Japanese would commence before their return live bombs were carried. During the flight, notification was received that hostilities had commenced and the Hudson diverted to Timor where it bombed and seriously damaged a 306 tonne Japanese vessel which was subsequently beached and abandoned. The destruction of this vessel was one of Japan's first maritime losses of the war.

After this first combat mission, reconnaissance and bombing operations were mounted at an ever increasing tempo. From January 1942 heavy losses to enemy fighters and anti-aircraft fire were being experienced. Nonetheless, 2 Squadron maintained its offensive effort and provided valuable information, particularly on Japanese shipping movements.

Despite the superiority of their aircraft and numerical advantage, the Japanese did not have everything their own way. On January 11, two Hudsons – captained by Flight Lieutenant P H R Hodge and Flying Officer P C Gorrie – were attacked by enemy floatplanes whilst they themselves bombed Japanese shipping. Although both Hudsons were damaged by return fire, two floatplanes were destroyed, one probably destroyed and a further one damaged.

The taste of victory experienced by these two gallant crews was short lived, however. Next day they, along with three other Hudson crews, resumed their bombing attacks against Japanese forces invading the Celebes. Caught by Zero fighters, two of the Hudsons were shot down before attacking and two others failed to return from the mission. Both Hodge's and Gorrie's crews were among those lost.

By mid February, 2 Squadron's advanced bases were under regular attack and the surviving aircraft of the detachments returned to Darwin on February 19. Any illusion the Hudson crews may have made about being at a safer locality was shattered when four more aircraft were destroyed on the ground during the first Japanese attack on Darwin which commenced just hours after they landed.

This and subsequent attacks forced a partial withdrawal to Daly Waters where bombing operations, principally against targets around Timor and the East Indies continued. In recognition of 2 Squadron's heroic stand in this, Australia's darkest hour, the unit was later awarded a United States Presidential Unit Citation – the highest honour that can be bestowed on a combat unit by the US Government.

2 Squadron moved to Batchelor in August and subsequently to Hughes in April 1943. In January 1944, the unit finally received its long awaited Hudson replacement when the first Beauforts were received. However a few months later, in May, the Beauforts were, in turn, exchanged for North American B-25 Mitchells.

With these well armed medium bombers, 2 Squadron specialised in anti-shipping strikes – achieving several notable successes – and daylight bombing attacks, primarily around Timor. While carrying out these operations, the Mitchells were regularly engaged by anti-aircraft defences and intercepted by Japanese fighters. Losses were relatively light and in return 2 Squadron inflicted substantial damage on enemy forces. A secondary role undertaken at this time was the aerial resupply of Australian troops on Timor – a task that was first conducted with Hudsons.

On April 6 1945, 2 Squadron provided ten Mitchells which participated in an unsuccessful attack against a Japanese convoy in the Flores Sea. Operations against the Japanese wound down later in the year in anticipation of a forward move. Unfortunately, these plans were frustrated by shipping shortages. After the cessation of hostilities, the Mitchells dropped supplies to

POWs – later evacuating some – conducted leaflets drops to the Japanese and made reconnaissance flights over Japanese occupied territory. Having completed these tasks, 2 Squadron, which had been operating from Balikpapan, returned to Australia where it disbanded at Laverton on May 15 1946.

On February 31 1948, 21 Squadron at Amberley was renumbered 2 Squadron. Operating Australian built Lincoln Mk.30s, it operated as a conventional bomber squadron participating in numerous exercises and civil aid tasks over the next several years until re-equipping with Canberra jet bombers in December 1953.

Apart from Amberley based routine training, 2 Squadron deployed its Canberras on regular exercises – sometimes overseas. Another unusual task undertaken by one crew was that of supporting the British atom bomb tests at Monte Bello in May/June 1956.

In July 1958, 2 Squadron moved to Butterworth on Malaya's west coast as part of the British Commonwealth Strategic Reserve. Here it provided an important element of Malaya's defence. 2 Squadron regularly participated in exercises with other Commonwealth air forces and, as the Malayan Emergency was not yet officially over, conducted several attacks against Communist Terrorist (CT) forces.

The first of these occurred in September when five Canberras struck a CT position in Northern Malaya. This strike was the RAAF's first jet bomber operation. Although results of this and four succeeding missions were difficult to assess, they undoubtedly contributed to the goal of destroying CT facilities and keeping them on the move.

After the Malayan Emergency had been concluded, 2 Squadron remained at Butterworth during the 1960s when tensions with Indonesia over that country's "Confrontation" policy with the newly formed Malaysia resulted in numerous armed incidents between Commonwealth and Indonesian forces. While held at a very high state of alert, it was not until 1967 that 2 Squadron was again to see action in the Vietnam War.

After much preparation, eight Canberras deployed to Phan Rang, South Vietnam, on April 19 1967. The unit was placed under the operational control of the United States 7th Air Force's 35th Tactical Fighter Wing (35th TFW) and on April 23 commenced operations against communist forces when all eight aircraft made ground radar directed attacks against enemy positions.

This first day's operations typified 2 Squadron's early flying in Vietnam – eight high level bombing attacks (code named "Combat Skyspots" were flown each 24 hours – predominantly by night after the crews had gained sufficient experience after a week of daylight attacks. To fly Combat Skyspots successfully, aircrews had to be extremely proficient instrument flyers as the radar controllers could give directions as little as ten feet in height and half a degree in course during the lead up to bomb release. Although immune from the threat of ground fire at the high altitudes these missions were flown, the Canberras were at times routed through severe tropical storms which could throw the aircraft up and down hundreds of feet where the crews often had little or no control over their aircraft.

After three months of constant night bombing, 2 Squadron began to experiment with daylight low level bombing, which appeared to be most promising. Initially, two missions per day were allocated to daylight attacks but as low level bombing proved more and more successful this was later increased to four and eventually eight missions. It is significant to note that 2 Squadron was the only tactical squadron in Vietnam to utilise this

From 1944 2 Squadron operated Mitchell medium bombers. (RAAF Museum)

A Canberra releases its bombs against a target in Vietnam. 2 Squadron gained a reputation as the most effective tactical bomber squadron in Vietnam for its precision attacks. (RAAF)

form of attack – USAF and South Vietnamese aircraft invariably made dive bombing attacks.

With its low level, daylight pinpoint strikes, 2 Squadron quickly established itself as the pre-eminent bomber squadron in Vietnam. Targets were usually obscured by jungle and the crews rarely could observe what they were attacking but, with the assistance of forward air controllers, the unit achieved remarkable results. While flying only five to six per cent of the 35th TFW's missions, it obtained some 16% of the assessed bomb damage.

So accurate were 2 Squadron's strikes, that 50% of the time, the Canberras attacked two separate targets on each mission and occasionally, three or four targets were destroyed during a single flight. 2 Squadron's excellence in air operations was matched by that of the ground staff who maintained a serviceability rate of 98% – the highest in the RAAF – for most of the unit's Vietnam tour.

On April 7 1971, as 2 Squadron was preparing to return to Australia, Flying Officer S Fenton and Pilot Officer P Murphy flew the unit's most successful mission of the Vietnam War when they were diverted from a preplanned strike to provide close air support to an American infantry company in heavy contact with North Vietnamese troops. Presuming themselves safe from air attack due to their proximity to the Americans, the North Vietnamese initially ignored the Canberra's presence until the first bomb landed in their midst. The Australian crew then made a further five runs over the target, despite heavy small arms fire being directed at them, releasing a single bomb on each occasion. The North Vietnamese force was systematically decimated and

those troops who survived fled into the jungle leaving eighty bodies on the battlefield.

On June 4 1971, 2 Squadron's Canberras departed from Phan Rang for Australia, with the ground staff following progressively over the succeeding weeks. During its Vietnam service, 2 Squadron flew 11,963 operational sorties and, for the loss of two aircraft, destroyed 7000 buildings, 10,000 bunkers, 1000 sampans, 36 bridges and an unspecified number of enemy troops killed. As a whole, the unit was awarded the Republic of Vietnam Cross of Gallantry and a United States Air Force Outstanding Unit Commendation. These two awards, combined with 2 Squadron's Presidential Unit Citation, made it the most highly decorated squadron in the RAAF.

After returning to Amberley, 2 Squadron maintained a limited bombing capacity for several years. The elderly Canberras were then confined to target towing for the RAAF's Mirage fighter squadrons – a task requiring regular deployments to Williamtown and Butterworth. Parallel to this commitment the unit also commenced photo survey operations in 1973. Survey tasks covered Papua New Guinea, Irian Jaya in Indonesia, large tracts of Australia, Cocos and Christmas Islands.

In 1979, serious stress corrosion cracking was found in many of 2 Squadron's Canberras and the difficulty in keeping the aircraft flying later forced the type's grounding in 1982. With no replacement aircraft forthcoming 2 Squadron was disbanded on July 31 1982.

Formed: *May 3 1937*
Squadron Codes: *B, KO*
Aircraft: *Demon, Anson, Wirraway, Hudson, Beaufort, Mitchell, Lincoln, Canberra*
Locations: *Laverton: May 1937 – Dec 1941*
Darwin: December 1941 – August 1942
Batchelor: August 1942 – April 1943
Hughes: April 1943 – August 1945
Balikpapan: August – December 1945
Laverton: December 1945 – May 1946
Disbanded:
Mallala: June 1947 – March 1948
Disbanded:
Amberley: February 1948 – July 1958
Butterworth: July 1958 – April 1967
Phan Rang: April 1967 – June 1971
Amberley: June 1971 – July 1982
Disbanded: *July 31 1982*

3 SQUADRON

3 Squadron formed on July 1 1925. Initially commanded by Flight Lieutenant F Barnes and based at Point Cook, the new unit, equipped with DH.9s, DH.9as and SE.5As, was classified a "composite" squadron with each of its three flights allotted separate roles – army co-operation, bombing and fighter.

A week after forming, 3 Squadron moved to Richmond and from its new base the unit, apart from strictly military flying, participated in a large number of air pageants and other public relations activities.

By 1930, 3 Squadron's conglomeration of World War I vintage aircraft had been replaced by a single type – the Westland Wapiti, however, within five years these were supplanted by Hawker Demon Is – one of the most elegant aircraft ever to serve with the RAAF.

After the commencement of World War II, 3 Squadron increased its flying rate markedly but it was not until 1940 that it was ordered overseas for active service. After handing over its aircraft and equipment to other units, 3 Squadron sailed for Egypt on July 15, arriving at Port Tewfik on August 23. The unit, which initially was to operate in the army co-operation role, moved to Ismailia and was equipped with Westland Lysanders and Gloster Gladiators and Gauntlets. It was then split up with the Lysanders deploying to Lkingi Mariut and the remaining aircraft

to Gerawla. Operations commenced on November 13 and the first air-to-air combat occurred on November 19 when four Gladiators were attacked by 18 Fiat CR.42 fighters. In the ensuing battle three Italian fighters were destroyed and another three probably destroyed for the loss of Squadron Leader P Heath who was shot down and killed.

3 Squadron continued its tactical reconnaissance and ground attack missions to the end of 1940, supporting the 8th Army in the ebb and flow of its desert campaign. By this time it was equipped solely with Gladiators and, operating from a number of airfields, regularly met and bested the Italian Air Force, despite being heavily outnumbered.

By January 1941, 3 Squadron received modern aircraft when it was allocated Hurricanes. On February 15 1941, the unit claimed its first "kill" against the Luftwaffe when Flying Officer J Saunders destroyed a Ju 88 bomber. From this time on air combat occurred with increased frequency and on February 18 three pilots destroyed eight Ju 87 Stuka dive bombers near Agedabia. Further successes followed and the unit's score of enemy aircraft increased rapidly despite the fact that, from April, it was forced to progressively withdraw due to the successful advance of German and Italian ground forces.

By May 1941, 3 Squadron, now operating from Lydda, re-equipped with Curtiss P-40 Tomahawks and in June began operations in support

Upon reforming after World War 2, 3 Squadron equipped with Mustangs and operated in the Army Co-operation role. (RAAF Museum)

of the Allied attack on Vichy French controlled Syria. Aside from flying tactical reconnaissance and close support missions, the unit successfully engaged French aircraft, gaining several more kills before the campaign was concluded in July.

On July 20, 3 Squadron moved to Rayak and rested until the commencement of the Second Libyan Campaign in November where the Tomahawks conducted fighter sweeps, bomber escort and ground attack missions intermingled with occasional tactical reconnaissance missions.

3 Squadron's score against enemy aircraft continued to mount despite the fact that the German Messerschmitt Bf 109s were superior to the Tomahawks in all round performance. On November 30, 10 enemy aircraft were destroyed, one of which was 3 Squadron's 100th kill. Having established itself as the Desert Air Force's pre-eminent fighter squadron, 3 Squadron was selected as the first unit in that Air Force to re-equip with Curtiss Kittyhawks, a process which began in December 1941.

By January 1942 the latest Allied offensive had failed and 3 Squadron found itself again retreating from airfield to airfield in the face of advancing German and Italian troops. With its ground echelon fully mobile and experienced in rapid movement, operations continued uninterrupted and the unit, flying intensively, supported the hard pressed ground forces and also met enemy aircraft in large numbers. One of the most spectacular engagements of this period occurred on February 14 when 18 Kittyhawks from 112 and 3 Squadron engaged 40 German aircraft, 20 of which were destroyed. 3 Squadron's share was eight and a half destroyed and five damaged.

On October 23, Squadron Leader R Gibbes shot down 3 Squadron's 200th enemy aircraft. This momentous event occurred in the middle of the Third Libyan Campaign which, by May 1943, resulted in the total defeat of German and Italian forces in Africa.

Australian built Sabres of 3 Squadron at Butterworth, Malaysia.

After a period at Zuara, 3 Squadron, operating from Malta, supported the invasion of Sicily until mid August when enemy forces on that island surrendered and the unit prepared for its next major operation – the provision of close air support during the invasion of Italy. As with the Sicilian campaign, the Kittyhawks flew intensively, mainly in the ground attack role. Losses from anti-aircraft fire were, at times, extremely heavy and in just one day four Kittyhawks were shot down – although three pilots survived.

On September 14, 3 Squadron moved to Grottaglie – its first base on the Italian mainland. Two weeks later the Kittyhawks moved northwards to the Foggia airfield complex, maintaining an intensive operational effort. By the end of October a further move, to Melini, had been effected and operations were conducted over a large part of Italy, Yugoslavia and surrounding waters.

Ground attack operations predominated into 1944 and very few enemy aircraft were encountered in the air. As German forces continued to withdraw up the Italian mainland, 3 Squadron moved from airfield to airfield to keep targets within range of its Kittyhawks. In this period the Australian aircraft gained special prominence for highly accurate attacks against enemy shipping and a considerable number of vessels were sunk or damaged.

3 Squadron received its first North American P-51 Mustangs in November and with these new aircraft continued its attack operations until May 2 1945 when German forces in Italy surrendered. The capitulation of Germany just five days later, precluded 3 Squadron deploying to another area of operations. Nonetheless, it had achieved remarkable results and aside from its vital ground attack operations, with a score of 217 enemy aircraft destroyed, 3 Squadron was the highest scoring fighter squadron of the Desert Air Force.

Sailing for Australia in September, 3 Squadron disbanded at Point Cook on July 30 1946.

Equipped with Mustangs, Austers and Wirraways, 3 Squadron reformed as a tactical

A 3 Squadron F/A-18 Hornet. (RAAF)

reconnaissance and close support squadron on March 8 1948. Based at Fairbairn, the unit engaged in routine training and exercises up until 1952 when the Mustangs and Wirraways were allocated to other units and 3 Squadron's role changed to the training of Army light aircraft pilots. On June 15 1953 it again disbanded.

Almost three years later, on March 1 1956, 3 Squadron rejoined the RAAF's order of battle when it reformed with Australian built CA-27 Sabre jet fighters at Williamtown. The unit trained both in the air combat and ground attack roles before deploying to Butterworth, Malaya, in October/November 1958. "Operation Sabre Ferry", the movement of 3 Squadron's Sabres to Butterworth was, in itself, a difficult and meticulously planned task with the short range jets staging through Darwin, Biak, Guian and Labuan.

Once settled into its new base, the unit trained much as it had in Australia, with the additional responsibility of providing part of Malaya's air defence. At this time the Malayan Emergency was in its final stages and 3 Squadron flew its first strike against Communist Terrorist forces on August 13 1959 when six Sabres in conjunction with another six from 77 Squadron bombed and strafed a terrorist camp. One or two other strikes were conducted prior to the Emergency being declared over in 1960.

3 Squadron remained at Butterworth for most of the turbulent 1960s when tensions in the region were stretched to breaking point over Indonesia's policy of "Confrontation" with the newly formed state of Malaysia. Acts of military aggression were regularly committed by Indonesian forces and for varying periods between 1963 and 1966 Sabres, armed with live ammunition and Sidewinder missiles, were held on

ground alert in the event of an Indonesian air attack. During this difficult period deployments were made to Singapore and Labuan.

Returning to Williamtown in February 1967, 3 Squadron re-equipped with the French designed, Australian built, Dassault Mirage III supersonic interceptor. From February 1969 the unit again deployed to Butterworth where regular air defence exercises were conducted as well as deployments to Tengah and later Paya Lebar at Singapore.

Preparatory to re-equipping with McDonnell Douglas F/A-18 Hornets, many of 3 Squadron's personnel and aircraft were transferred to the newly formed 79 Squadron at Butterworth during March 1986. At the same time the unit ceased to exist at Butterworth and was re-established at Williamtown. The first Hornet was received on August 29 and the unit slowly worked up to operational status.

Since that time 3 Squadron, still at Williamtown, has participated in regular air defence exercises, both in Australia and as far a field as Malaysia, Singapore, the Philippines and New Zealand. With its missile armed, long range multirole Hornets, 3 Squadron will maintain a preeminent place as one of Australia's most vital defence assets for decades to come.

Formed: *July 1 1925*
Squadron Code: *C, NW, CV*
Aircraft: *DH.9, DH.9a, SE.5A, Wapiti, Demon, Lysander, Gladiator, Gauntlet, Hurricane, Tomahawk, Kittyhawk, Mustang, Sabre, Mirage, Hornet*
Locations: *Point Cook: July 1925 – July 1925*
Richmond: July 1925 – July 1940
Helwan: September – November 1940

Gerawla: November 1940 – January 1941
Tmimi: January 1941
Martuba: January – February 1941
Berka: February 1941
Benina: February 1941
Got-es-Sultan: April 1941
Maraua: April 1941
Martuba: April 1941
Gazala East: April 1941
Sidi Mahmoud: April 1941
LG 79: April 1941
Mersa Matruh: April 1941
Sidi Haneish: April 1941
Aboukir: April – May 1941
Lydda: May – July 1941
Rosh Pinna: July 1941
Rayak: July – September 1941
Amriya: September 1941
Sidi Haneish: September – November 1941
LG 110: November 1941 – February 1942
Sidi Haneish: February – June 1942
Wadi Natrun: June 1942
LG 91: June 1942
Amriya: June – July 1942
LG222: July – September 1942
LG 91: September – November 1942
LG 106: November 1942
LG 101: November 1942
LG 76: November 1942
Gambut: November 1942
Gazala: November 1942
Martuba I: November – December 1942
Antelat: December 1942
Belandah: December 1942
Marble Arch: December 1942
Adem el: December 1942
Gzina: December 1942 – January 1943

Hamraiet: January 1943
Sedada: January 1943
Castel Benito: January – February 1943
El Assa: February – March 1943
Nefatia: March 1943
Medenine Main: March – April 1943
El Hamma: April 1943
Kairouan: April – May 1943
Zuara: May – July 1943
St Pauls Bay: July 1943
Takali: July 1943
Luqu: July 1943
Pachino: July – August 1943
Agnone: August – September 1943
Grottaglie: September 1943
Bari: September – October 1943
Foggia Main: October 1943
Mileni: October 1943 – January 1944
Cutella: January – May 1944
San Angelo: May – June 1944
Guidonia: June 1944
Falerium: June – July 1944
Crete: July – August 1944
Iesi: August – November 1944
Fano: November 1945 – February 1945
Cervia: February – May 1945
Lavariano: May – August 1945
Point Cook: August 1945 – July 1946
Disbanded:
Fairbairn: March 1948 – January 1953
Disbanded:
Williamtown: March 1956 – November 1958
Butterworth: October 1958 – February 1967
Williamtown: February – 1967 – February 1969
Butterworth: February 1969 – March 1986
Williamtown: March 1986 – present

4 SQUADRON

4 Squadron formed at Richmond on May 3 1937 initially with Hawker Demons and Avro Ansons. However, after a short existence of less than two years, the unit, commanded by Squadron Leader D E L Wilson, was renumbered 6 Squadron.

Prior to this occurring, 4 Squadron regularly conducted seaward patrols and participated in a number of airshows and other public events. One important activity undertaken in this period occurred during September 1938 when an Anson was sent to Northern Australia to search for Dr Fenton – the Flying Doctor – whose aircraft went missing during one of his many humanitarian flights.

4 Squadron again came into existence at

Richmond on June 17 1940 equipped with Hawker Demon, de Havilland Moth Minor and later CAC Wirraways. The unit trained in a variety of roles including dive bombing, ground attack and photo reconnaissance.

This training, intermingled with exercises and other activities, continued after 4 Squadron relocated to Camden in July 1942. A further move to Kingaroy was effected during September, prior to the unit being allocated to the New Guinea area for combat operations.

This long awaited movement occurred in November 1942 when the Wirraways were established at Port Moresby's Berry strip. From here 4 Squadron supported Australian troops fighting

4 Squadron Wirraway. (RAAF Museum)

desperately against Japanese forces. The Wirraways, often operating as detachments away from Port Moresby, flew a variety of missions which included artillery spotting, supply and message dropping, strafing and bombing and reconnaissance.

In their slow and vulnerable aircraft, losses to enemy fire were constant, however, this never deterred the Wirraway crews from completing their assigned tasks. This aggressive spirit was exemplified by Pilot Officer J S Archer and his crewman, Sergeant J L Coulston who, whilst on a tactical reconnaissance mission near Buna on December 26, found themselves above a Japanese Zero fighter. Despite operating a vastly inferior aircraft, Pilot Officer Archer dived to the attack and shot down the enemy aircraft. This was the only occasion where a Wirraway shot down an enemy aircraft in World War II.

This incident well illustrated the versatility and skill which was a regular feature of 4 Squadron's work. On some missions the Australian airmen provided support for American ground forces. The successful nature of this support resulted in a number of pilots being awarded United States gallantry awards – principally Silver and Bronze Stars. A typical example was Flying Officer J R Mobray who received a Silver Star for "making a human target of himself in order to confirm an anti-aircraft gun's presence by making it fire and so disclose its position". So typical were actions such as these that United States General Eichelberger was moved to write of the Australian Wirraway crews "I can hope to serve with no braver men".

In June 1943, 4 Squadron received its first CAC Boomerangs. These new aircraft were not popular as the Wirraways had performed well in their assigned role. The Boomerang, also was similar in appearance to several enemy aircraft types and this was tragically highlighted when one was shot down by friendly anti-aircraft fire over Nassau Bay, killing the pilot, Flying Officer J Collier.

Later, another Boomerang was badly damaged when attacked by American fighters in error. Despite the advent of the Boomerang, two or three Wirraways were retained for use on those tasks in which two seat aircraft were more suited – particularly artillery spotting and supply dropping.

4 Squadron records dispel the inaccurate contention that the Boomerang never engaged in air-to-air combat but only in the most tragic circumstances.

On November 26 1943, two Boomerangs, flown by Flying Officer Munro and Flight Sergeant Slater, were conducting a tactical reconnaissance when they were engaged by seven Japanese fighters over the Sanga River. Eyewitnesses saw the Australian aircraft engaging the enemy fighters, however, neither Boomerang returned to base and it was assumed that both had been shot down.

From March 1944, 4 Squadron supported Allied troops during the Cape Gloucester landings before moving to Moratai. 4 Squadron ended the war in Borneo where it flew in support of Australian troops engaged in large scale mopping up operations against bypassed Japanese forces.

By January 1946, 4 Squadron had moved to Fairbairn and was reduced to a nucleus with very few personnel and only two Kittyhawks. In early 1947 the unit was allocated a tactical reconnaissance role and received its first Mustangs and Austers. Personnel numbers were also built up. As had occurred earlier in 4 Squadron's history, however, the unit was again renumbered, this time as 3 Squadron on March 8 1948.

Formed: *May 3 1937*
Squadron Code: *D, QE*
Aircraft: *Demon, Moth Minor, Anson, Wirraway, Boomerang, Kittyhawk, Mustang, Auster*
Locations: *Richmond: May 1937 – Jan 1939*
Disbanded:
Richmond: June – November 1940
Canberra: November 1940 – May 1942
Camden: May – September 1942
Kingaroy: September – November 1942
Port Moresby: November 1942 – April 1945
Moratai: April – June 1945
Labuan: June 1945 – January 1946
Canberra: January 1946 – March 1948
Disbanded: *March 8 1948*

---

5 SQUADRON

5 Squadron formed at Richmond under the command of Flight Lieutenant L V Lachal on April 20 1936. Equipped with Supermarine Seagull V amphibians and personnel inherited from 101 (Fleet Co-operation) Flight, the unit operated from RAN cruisers and briefly from the seaplane carrier HMAS *Albatross*, providing the Navy with a reconnaissance and artillery spotting capability. Survey operations, in conjunction with HMAS *Moresby* were also conducted before the unit was renumbered 9 Squadron on January 1 1939.

Equipped with Australian built CAC Wirraways, 5 Squadron reformed at Laverton on January 9 1941. As opposed to its former naval co-operation tasks, the new 5 Squadron's role was that of army co-operation. The unit trained in tactical and photo reconnaissance, artillery spotting, message dropping and other army support tasks. These exercises often involved the detachment of aircraft away from Laverton and continued after the outbreak of war.

In May 1942, 5 Squadron moved to Toowoomba and succeeding moves took the unit to Toogoolawah (November), Kingaroy (February 1943) and Mareeba (June 1943). In October 1943, 5 Squadron received its first Australian built CAC Boomerangs – an aircraft which largely, but not entirely replaced its Wirraways. Despite the allo-

cation of these new aircraft, it was not until late in the war that 5 Squadron was to see active service.

On November 11 1944, 5 Squadron's Boomerangs and Wirraways flew to Piva North on Bougainville and the next day the first combat mission – a tactical reconnaissance – was flown. Artillery spotting, tactical reconnaissance and close support missions in support of Australian troops continued up to the end of the war.

One important task undertaken was to identify Japanese positions and then lead in RNZAF Corsairs on their attack runs. To effectively fulfil these tasks required considerable risks on the part of the Australian airmen who had to fly at low level constantly exposed to the threat of ground fire.

From February 1945, a detachment of Boomerangs and a Wirraway began operations from Cape Hoskins and later Tadji on the New Guinea mainland. From mid August the unit dropped leaflets on Japanese positions, advising the enemy that hostilities had ceased.

In September, 5 Squadron took delivery of one or two Kittyhawks, however, flying commitments progressively declined towards the end of the year and the unit was reduced to cadre status prior to relocating to Pearce in February 1946. Here it began to re-equip with Kittyhawks but plans to maintain the unit in the early postwar years lapsed and 5 Squadron disbanded on October 18 1946.

Manufactured by the Commonwealth Aircraft Corporation, the Boomerangs of 5 Squadron proved highly effective in the Army co-operation role. (RAAF)

5 Squadron's last aircraft type – the French built Squirrel. (Glenn Alderton, RAAF)

Equipped with Bell UH-1B Iroquois helicopters, 5 Squadron reformed at Fairbairn on May 4 1964 and just a few weeks later, on June 13, deployed to Butterworth, Malaysia, where it was to support Malaysian and British Commonwealth troops hunting Indonesian infiltrators during the "Confrontation".

After settling into its new base, the Iroquois were used to insert and extract troops into remote areas and then resupply them with food and ammunition. Other roles included reconnaissance and medical evacuation of casualties. In the thick jungle which covered much of Malaysia, the mobility and operational flexibility provided by the helicopters was especially valuable to the ground forces. The Australian helicopter crews also gained valuable experience of jungle operations – experience to be utilised in another war zone just two years later. Aside from their operational work, the utility of 5 Squadron's Iroquois ensured that they were also used regularly in support of the civil community.

On April 12 1966, that portion of 9 Squadron at Fairbairn not deploying on active service in Vietnam became 5 Squadron. The former 9 Squadron search and rescue detachments at Williamtown and Darwin, along with the original 5 Squadron at Butterworth became detached sub-units of the parent squadron at Fairbairn. With the Indonesian conspired "Confrontation" against Malaysia rapidly decreasing in tempo, 5 Squadron's operations in Malaysia reduced in scale and the detachment ceased to function in May 1966.

Meanwhile, the Williamtown and Darwin detachments continued to maintain their search and rescue capacity while the bulk of the unit was heavily committed to supporting infantry battalions training for Vietnam. For the duration of the Vietnam War – and for many years afterwards – 5 Squadron was also responsible for conversion training of all RAAF helicopter crews. Apart from these important roles, the Fairbairn based helicopters were used extensively in flood relief work, searches for lost people and various other civil aid tasks.

Scheduled operations were disrupted in July 1976 when four helicopters, maintenance personnel and aircrews were deployed to Ismailia, Egypt for service with a United Nations peacekeeping force whose task was to monitor a ceasefire between Arab and Israeli forces.

Aerospatiale AS 350B Squirrels arrived at Fairbairn in May 1984 to replace the by now elderly B model Iroquois used in pilot training. With these new helicopters, and the H model Iroquois which soldiered on in the army support role, 5 Squadron continued its activities until December 9 1989 when it disbanded and was absorbed into the Australian Defence Force Helicopter School.

Formed: *April 20 1936*
Squadron Code: *E, BF*
Aircraft: *Seagull, Wirraway, Boomerang, Kittyhawk, Iroquois, Squirrel*
Locations: *Richmond: April 1936 – Jan 1939*
Disbanded:
Laverton: January 1941 – May 1942
Toowoomba: May – November 1942
Toogoolawah: November 1942 – February 1943
Kingaroy: February – June 1943
Mareeba: June 1943 – November 1944
Piva North: November 1944 – February 1946
Pearce: February – October 1946
Disbanded:
Fairbairn: May – June 1964
Butterworth: June 1964 – April 1966
Fairbairn: April 1966 – December 1989
Disbanded: *December 9 1989*

6 SQUADRON

On January 1 1939, 4 Squadron at Richmond was renumbered 6 Squadron. Commanded by Wing Commander D E L Wilson, the unit's Anson Mk.I aircraft were used for general reconnaissance. Shortly after its formation, the training of general reconnaissance aircrews also became a responsibility of the unit.

6 Squadron's first warlike operation was flown on August 26 when eight Ansons searched fruitlessly for the German freighter *Lehne* which had sailed illegally from Sydney Harbour the previous evening. April 1940 saw the unit begin to re-equip with Lockheed Hudsons and Japan's entry in the war during December 1941 resulted in an increased number of anti-submarine and general reconnaissance patrols being flown.

In January 1942, two especially modified Hudsons departed Richmond and, staging through Townsville and various island bases, conducted the longest photographic reconnaissance mission undertaken by land based RAAF aircraft during World War II. One Hudson became unserviceable, however, the second aircraft, captained by Flight Lieutenant R Yeowart, ignored anti-aircraft fire, and Japanese fighters which attempted interception, and photographed Truk harbour and airfield, obtaining intelligence information of considerable importance.

6 Squadron moved to Horn Island in August 1942 and, almost immediately, a flight of four Hudsons and a maintenance detachment deployed to Milne Bay to provide reconnaissance and bomber support to the Australian garrison which daily faced the prospect of being attacked by superior Japanese forces.

6 Squadron operated Australian built Beauforts in the last two years of the Pacific war.

The Japanese made their expected landing on the night of August 25/26 and, during the following days, the Hudson crews flew intensively, bombing and strafing landing barges, stores and troops in the bridgehead and any ships located in daylight. These attacks, which complemented those of two RAAF Kittyhawk squadrons, were particularly damaging and resulted in considerable loss to the Japanese.

On August 29, Squadron Leader D Colquhoun's crew located a Japanese cruiser and eight destroyers near Normanby Island. After reporting the ship's position the Hudson made a daring lone attack through a curtain of anti-aircraft fire. One bomb detonated under the stern of a destroyer, lifting the ship partially out of the water and severely damaging it.

After some initial success on the ground, the Japanese ground troops were fought to a standstill by a stubborn Australian ground defence supplemented by extremely effective air attacks. Realising that their plans were doomed to failure, the Japanese evacuated Milne Bay during the second week of September in what was their first unquestionable defeat on land. Milne Bay also represented the most southerly point of Japan's advance during World War II.

After the Japanese withdrawal, reconnaissance missions continued from Milne Bay. A number of attacks resulted from these flights and several ships were sunk or damaged. On September 26, three Hudsons located and attacked a transport ship off Woodlark Island. The third aircraft scored two direct hits which blew the vessel's bridge off and caused so much damage that it sank soon afterwards.

In October, that portion of 6 Squadron operat-

Having crashed on landing at Amberley this 6 Squadron Phantom, despite major structural damage, was later repaired. The USAF commented that the repaired F-4 was returned in a better condition than when it left the factory new.

ing from Horn Island relocated to Port Moresby where the Hudsons dropped stores to Australian troops fighting their way across the rugged Owen Stanley Ranges along the Kokoda Trail. Although not an ideal aircraft for this task some 23,744 kilograms of supplies were delivered under extremely hazardous conditions. October saw a further maritime success achieved when a 10,200 tonne vessel with a deck cargo of 12 Zero fighters was attacked and left burning in the St Georges Channel by a patrolling Hudson.

By November 1942, the entire unit was located at Milne Bay from where reconnaissance and attack missions were continued. On the night of January 17 1943, 6 Squadron was heavily hit in a Japanese reprisal air raid which saw one Hudson destroyed and all the remaining aircraft suffering damage. Fortunately, the war worn Hudsons were repaired but, in August, were replaced by Australian built Bristol Beauforts.

On October 8, Flying Officer W A Barr's Beaufort was attacked by six Japanese fighters while on a reconnaissance flight. In a remarkable engagement, two of the fighters were probably destroyed, one damaged and the other three driven off while the slow and poorly armed bomber sustained only minor damage.

Three days after this incident, Japanese float planes attacked Flying Officer Hale's Beaufort in the St Georges Channel. Evading these aircraft, the Beaufort crew located a Japanese convoy and bombed a 2040 tonne ship which was set on fire with a direct hit. Flying Officer Hale's crew were again prominent in a shipping strike on October 20 when attacking a well defended convoy. Despite heavy anti-aircraft, fire Flying Officer

Hale's bomb aimer scored a direct hit on a cruiser – the bomb apparently dropping down the vessel's funnel!

By November 1943, 6 Squadron had moved to Goodenough Island where many night strikes against the Japanese citadel at Rabaul were flown. These attacks, often pressed home through severe tropical storms, not to mention heavy anti-aircraft fire, searchlight and sometimes night fighter defences, were a vital component of a combined American/Australian round the clock bombing campaign designed to destroy Rabaul's air and naval forces, and to isolate its garrison of over 100,000 troops.

In December 1944, 6 Squadron relocated to Dobodura on the New Guinea mainland where strikes and supply dropping missions in support of Australian troops on New Britain were flown. By early 1945, however, operations had tapered off to the extent that, with no worthwhile targets remaining within range, the unit was virtually non-operational for the remainder of the war. The only real combat operations flown in this period occurred when a flight of Beauforts joined other squadrons at Tadji which were supporting the Australian Army in its drive to Wewak. After returning to Australia at war's end, 6 Squadron disbanded at Kingaroy on October 31 1945.

One of the most elegant aircraft operated by the RAAF – the Learjet. 6 Squadron operated leased Learjets in the photo survey role for several years during the 1980s.

A 6 Squadron F-111G. (Keith Anderson)

On February 23 1948, 23 Squadron, equipped with Avro Lincoln Mk.30 bombers at Amberley, was renumbered 6 Squadron. As part of 82 Bomber Wing the new 6 Squadron operated primarily as a training unit, providing aircrews for eventual service with both 1 and 2 Squadrons. With the deployment of 1 Squadron to participate in the Malayan Emergency during 1950, this training effort was considerably stepped up.

Aside from its important training role, 6 Squadron undertook civil aid tasks such as flood relief, bushfire patrols and searches for missing vessels. The unit also had the misfortune to participate in the British atom bomb tests at Monte Bello Island and later, Maralinga. During these tests the Lincolns monitored the progress of the atomic clouds by flying through them with geiger counters fitted to the aircraft. Some of the Lincolns were so heavily contaminated by this activity that they were never flown again and were later destroyed. Little has been said on the effects of radiation upon the crews.

The date of April 8 1953 is one remembered with considerable embarrassment by 6 Squadron – and with good reason. During the morning Flight Lieutenant S W Trewin was landing at Amberley when his Lincoln's undercarriage collapsed without warning. Meanwhile, Flight Lieutenant K Isaac and crew were landing at Cloncurry during a navigation exercise. On touchdown the aircraft sustained a metal fracture which forced the starboard undercarriage to collapse. That evening, Flight Lieutenant Winchcombe left Amberley with spare parts for the damaged Lincoln at Cloncurry. Unfortunately, while landing in the dark the Lincoln crashed into its damaged compatriot, totally destroying both aircraft. In this remarkable series of incidents the aircrews involved escaped injury, however, all three bombers were written off.

July 11 1955 saw the commencement of Canberra jet operations. Routine training continued until early 1967 when 6 Squadron was virtually absorbed by 2 Squadron to bring that unit up to its war establishment prior to deploying to Vietnam. Once a skeleton, 6 Squadron had been brought back up to strength training aircrews for combat operations in Vietnam. This proceeded at high pitch until 1970 when plans to re-equip 6 Squadron with F-4E Phantom fighter bombers bore fruit. The delivery of Phantoms commenced in September and was completed the following month.

With their excellent performance and the quantum leap in technology they represented over the Canberras, the supersonic Phantoms proved popular with air and ground crews, not to mention being great crowd pleasers at the RAAF's Golden Jubilee Airshows held during 1971. The Phantom's combat capability was so great that the aircrews found that they no longer required fighter escort even during the most demanding of air defence exercises.

The Phantoms had only been leased as a interim measure however, and after a final formation flypast on October 4 1972 operations ceased preparatory to the delivery of long awaited swing-wing F-111C strike aircraft. Despite being excited about the impending arrival of their new charges, personnel were distinctly unenthusiastic about returning their McDonnell Douglas built jets.

The first General Dynamics F-111Cs arrived in Australia on June 1 1973 and, with subsequent deliveries, 6 Squadron was brought up to full strength with these sophisticated all weather long range attack aircraft. Apart from strike operations, 6 Squadron was given the task of aircrew conversion training for itself and 1 Squadron. A new capability was added in August 1979 when the first of four F-111Cs were converted for photo reconnaissance operations and redesignated RF-111Cs.

With the allocation of leased Learjet aircraft from June 1982 photo survey operations in conjunction with the Army Survey Corps commenced. The Learjet section undertook major survey tasks within Australian and overseas until financial constraints forced its disbandment in 1987.

Over recent years, 6 Squadron has participated in numerous international competitions and exercises both in the strike and reconnaissance roles. These deployments have established the unit as one of the most capable attack/reconnaissance squadrons in the world. 6 Squadron continues to operate F-111C and RF-111C aircraft – supplemented since 1993 by F-111Gs – from Amberley and, with a planned avionics update well underway, the F-111 is expected to remain in service until the year 2010 and perhaps longer.

Formed: *January 1 1939*
Squadron Code: *F, FX*
Aircraft types: *Anson, Hudson, Beaufort, Lincoln, Canberra, Phantom, F-111A/C/G, RF-111C, Learjet*
Locations: *Richmond Jan 1939 – Aug 1942*
Horn Island: August – October 1942
Port Moresby: October 1942 – January 1943
Milne Bay: January – November 1943
Goodenough Island: Nov 1943 – Feb 1945
Dobodura: February – October 1945
Kingaroy: October 1945
Disbanded:
Amberley: February 1948 – present

7 SQUADRON

Under the command of Squadron Leader E D Scott, 7 Squadron formed at Laverton on June 27 1940. It was intended to arm the Squadron with Hudsons, however, none were allotted and most personnel were eventually posted away or attached to 2 Squadron, leaving the unit as little more than a cadre.

7 Squadron again began to form in January 1942, however, its role was changed from that of a bomber squadron to that of a temporary Hudson operational training unit, although some anti-submarine patrols and convoy escorts were also flown.

In April, 7 Squadron moved to Bairnsdale with a detached flight at Mallacoota. Only weeks later, on June 4 a Hudson captained by Flight Lieutenant C Williams attacked and possibly damaged a Japanese submarine. June saw 1 Operational Training unit move to Bairnsdale and, after absorbing most of 7 Squadron, only a small nucleus remained to move to Nowra in August for rearming with Beauforts. Conversion training was completed in October and 7 Squadron moved to Ross River where it was to operate in the bomber-reconnaissance role.

From Ross River, patrols over Australia's northern waters were undertaken. Generally

Beaufort of 7 Squadron. (RAAF Museum)

these were uneventful but on December 15 1942 a Japanese submarine was caught on the surface and probably damaged by Flying Officer Whitshaw's crew. In at least five other incidents Japanese Jake floatplanes – detected on the Beaufort's radar – were intercepted and engaged by the Australian crews. Two of these were shot down and some of the others damaged.

After spending six months on Horn Island, 7 Squadron moved to Tadji in October 1944 where attacks against Japanese positions were made in support of Australian troops. The first of these strikes occurred on November 27 1944 when five Beauforts made low level bombing and strafing attacks against targets on Merauke. Whilst at Tadji, a number of supply dropping missions were also flown.

7 Squadron's last mission of the war was flown on August 15 1945 when, hours before Japan's surrender, 12 aircraft struck targets in the Maprik area. For several days after the cessation of hostilities, surrender leaflets were dropped to the Japanese. On December 19 1945, 7 Squadron disbanded at Tadji.

Formed: *January 27 1942*
Squadron Code: *G, KT*
Aircraft: *Hudson, Beaufort*
Locations: *Laverton: June 1940 – April 1942*
Bairnsdale: April – August 1942
Nowra: August – October 1942
Ross River: October 1942 – April 1944
Horn Island: April – October 1944
Tadji: October 1944 – December 1945
Disbanded: *December 19 1945*

---

8 SQUADRON

Initially equipped with Douglas DC-2s and DC-3s for want of more suitable reconnaissance aircraft, 8 Squadron under the command of Squadron Leader R H Simms, formed at Fairbairn on September 11 1939.

8 Squadron conducted seaward searches, various patrols and transport tasks until re-equipped with Hudsons in May 1940. In the latter part of the year, the unit moved to Sembawang, Singapore, and from this tropical base participated in bombing, reconnaissance and army co-operation exercises until hostilities with Japan commenced on December 8 1941.

That fateful day found 8 Squadron unpacking after having relocated to Kuantan one day previously. Despite confusion caused by the move, combat operations against Japanese shipping off Kota Baharu were mounted that morning. Twelve Hudsons were despatched on the first strike of the day and, despite fighter opposition and anti-aircraft fire, made effective attacks against the Japanese bridgehead. Enemy action during this initial strike resulted in two bombers making forced landings while three others suffered major damage. In return, and aside from casualties inflicted on their ground forces, one Japanese fighter was known to have been shot down by Sergeant R S Jansen, a gunner in Flight Lieutenant Hitchcock's Hudson.

After conducting further bombing and reconnaissance missions, 8 Squadron returned to Sembawang on December 9 after Japanese aircraft raided Kuantan, destroying base facilities and three Hudsons on the ground. Due to a lack of replacement aircraft, 8 Squadron was virtually amalgamated with 1 Squadron at Sembawang, the combined squadron conducting seaward searches and patrols over the South China Sea despite strong fighter opposition. These reconnaissance missions were so routinely intercepted that they became known as the "clay pigeon run". An additional problem at this time was the damaging attacks mounted against Sembawang which commenced soon after the Hudsons returned there.

Having transferred its few surviving aircraft to 1 Squadron towards the end of January 1942, 8 Squadron moved to Palembang II airfield on Java. Here replacement Hudsons were used to conduct vital reconnaissance and attack missions until Java was invaded in the second week of February. For the next few days, strikes were maintained against the Japanese until, with aircraft numbers dwindling, the unit was evacuated to Batavia on February 16. Four days later, 8 Squadron was disbanded and its personnel returned to Australia to avoid capture.

On March 12 1943, 8 Squadron reformed at Fairbairn with Australian built Bristol Beauforts prior to relocating to Bohle River. In August, the unit, which specialised in torpedo and level bombing roles, moved to Goodenough Island and from here, commenced night attacks against shipping and various land targets in the New Britain area – in particular Rabaul.

While not overly successful in the torpedo

Reforming after its participation in the Malayan campaign, 8 Squadron was allocated Australian built Beauforts.

bombing role, a small force of Beauforts gained a notable success on the night of December 4 when a convoy was attacked in the approaches of Rabaul's Simpson Harbour. Despite heavy anti-aircraft fire which shot down the Beaufort captained by Squadron Leader N T Quinn, one 6834 tonne ship was torpedoed and sunk.

With the Japanese forces at Rabaul effectively neutralised, 8 Squadron moved to Nadzab in April 1944 where attacks were concentrated on the But, Wewak and Borum areas. A further move, to Tadji, was made in mid June. From here the Beauforts supported the bitterly contested Australian ground offensive towards Wewak with numerous bombing attacks against enemy positions. As a secondary role, anti-submarine patrols were also conducted. Flying activity rapidly decreased after the Japanese surrender and 8 Squadron disbanded at Tadji on January 19 1946.

Formed: *September 11 1939*
Squadron Code: *H, NN, UV*
Aircraft: *DC-2, DC-3, Hudson, Beaufort*
Locations: *Fairbairn: Sept 1939 – August 1940*
Sembawang: August 1940 – February 1941
Kota Baharu: February – August 1941
Sembawang: August – December 1941
Kauntan: December 1941
Sembawang: December 1941 – January 1942
Palembang: January – February 1942
Batavia: February 1942
Disbanded:
Fairbairn: March – July 1943
Bohle River: July – August 1943
Goodenough Island: August 1943 – April 1944
Nadzab: April – June 1944
Tadji: June 1944 – January 1946
Disbanded: *January 19 1946*

9 SQUADRON

On January 1 1939, 5 Squadron, based at Richmond, was renumbered 9 Squadron.

The unit, equipped with Supermarine Seagull V amphibians (known as the Walrus in Royal Air Force service), was destined to be the RAAF's only naval co-operation squadron during World War II.

The Seagulls were designed to be embarked on Navy cruisers – and later armed merchant cruisers – to provide reconnaissance, artillery spotting and general support for the Navy. Based from separate ships and with such a varied role, 9 Squadron operations typically involved indi-

vidual aircraft flying in areas often on the other side of the world from Richmond. When at sea the aircraft and its air and ground crews came under the operational control of the captain in whose ship they served.

During the lead up to World War II the Seagulls participated in most naval exercises and proved a most versatile aircraft in their intended role. The Seagulls were catapulted from their parent ship and after completing their assigned mission, landed on the sea to be picked up by crane and secured until required next.

After the declaration of war, several Seagulls

and their crews, embarked in their parent cruisers, found themselves in Mediterranean waters where they served for varying periods until early 1942. Other aircraft remained in Australian waters and while operating off the Western Australia coast in December 1940, HMAS *Canberra's* Seagull conducted search flights for German surface raiders. Although unsuccessful, the amphibian enabled *Canberra* to search a much larger area than would otherwise have been the case. Another Seagull, captained by Flying Officer L G Webber was embarked in the New Zealand cruiser HMNZS *Achilles* and operated extensively throughout the Pacific with that ship.

In late 1940 HMAS *Australia's* Seagull was replaced by a Walrus and this aircraft, with its RAAF pilot and RAN crew, was used to search for enemy shipping around Bear Island. Not long after it directed a bombardment by *Australia* during an operation against Vichy French forces at Dakar. While so engaged it was set upon by two French fighters. The biplane stood no chance and was shot down into the sea with no survivors.

The light cruiser HMAS *Hobart's* amphibian was also put to good use in the Mediterranean. Apart from routine reconnaissance and patrol duties, the small aircraft made bombing attacks on targets on Centre Peak Island in the Red Sea and also against the Italian airfield at Berbera. In the latter attack some damage was inflicted on facilities with the aircraft's small bomb load while in return, the amphibian was holed by anti-aircraft fire.

HMAS *Perth's* detachment also had an interesting, although chequered, tour in the Mediterranean. In July 1940 the ship's aircraft's wings collapsed from the blast effect of *Perth's* main armament while, in January 1941, the amphibian

was strafed and badly damaged at Heraklion, Greece. Prior to this incident the aircraft had successfully completed reconnaissance and communication flights around Greece. Later, on April 28, a replacement Walrus captained by Flight Lieutenant E V Beaumont was engaged by two German bombers and after a plucky fight was forced down into the sea where its crew was later rescued by a passing destroyer.

The Seagull detachment in HMAS *Sydney* also completed good work but again suffered its own share of misadventures on June 21 1941 while directing naval gunfire against the Italian fortress town of Bardia. The amphibian, captain by Flight Lieutenant T McBride-Price was attacked in error by friendly fighters and was so badly damaged that it was forced to crashland at Matruh.

With the entry of Japan into the war in December 1942, the Australian cruisers returned to the Pacific theatre where they could play a direct role in the defence of Australia. During the crucial Guadalcanal campaign *Australia's* aircraft, in conjunction with US Navy float planes, conducted anti-submarine patrols in support of Allied naval forces.

9 Squadron members also suffered the fate of their ships companies and when HMAS *Canberra* was sunk in a night action off Guadalcanal on August 8 1942 five 9 Squadron personnel were killed and a further two wounded. Similarly, when *Sydney* was lost with all hands in an action with the German surface raider *Kormoran* on November 19 1941 six RAAF members were among those killed. Another five personnel were lost on March 1 1942 when *Perth* was sunk in a gallant action against a vastly superior Japanese naval force in the Sundra Strait.

For 9 Squadron, however, operations in the Pacific theatre from 1943 onwards resulted in a much less important role. The RAN was operating closely with the United States Navy – a force

A Seagull of 9 Squadron alights. (RAAF Museum)

which had available an ever increasing number of aircraft carriers. Consequently, the role of the cruiser borne amphibian aircraft had generally become redundant. By late 1944, all the RAN cruisers had their aircraft and catapults removed and, as a result, no effective role remained for 9 Squadron to fulfil and the unit disbanded at Rathmines on December 31 1944.

Allocated a search and rescue role, 9 Squadron reformed at Williamtown on June 11 1962. Personnel were sent to the United States for training on the Bell UH-1B Iroquois helicopters which had been ordered for the unit. After moving to Fairbairn in November, 9 Squadron received its first Iroquois and, as further helicopters were delivered, 9 Squadron increased its training effort and quickly became operational.

A 9 Squadron gunner replies to Viet Cong fire in Phuoc Tuy Province.

From early 1963, 9 Squadron began search and rescue operations, both for the RAAF and civilian authorities. Experimental army support tasks were also conducted and proved so successful that this type of operation quickly began to predominate over search and rescue. 9 Squadron continued peacetime flying – often co-operating with troops who would shortly be deployed to Vietnam – for the next few years until itself being committed to Vietnam in mid 1966.

Established at Vung Tau, South Vietnam, by June 12 1966, 9 Squadron flew its first mission the next day when two helicopters delivered ammunition to troops of the 5th Battalion. Apart from ammunition and food resupplies, roles for the RAAF helicopters included troop insertions and extractions and later medical evacuations, known as Dust Offs in Vietnam.

During the battle of Long Tan on August 18 1966, two Iroquois captained by Flight Lieutenants C Dohle and F Riley successfully completed an extremely hazardous resupply mission to the heavily outnumbered Australian and New Zealand ground troops. This resupply proved to be a critical factor in the outcome of the battle. After the enemy had withdrawn, several helicopters evacuated casualties from the battlefield.

9 Squadron lost its first helicopter on October 18 when an Iroquois hit trees and caught fire while landing in an extremely small landing zone. Sergeant G Buttriss, despite being injured himself, dragged his crewmates from the wreckage and they were later rescued.

By late 1967 it was realised that 9 Squadron would require substantial reinforcement if it were to fulfil its obligations to the growing Aus-

tralian force in Vietnam. Consequently, its establishment was increased from eight to 16 helicopters. At the same time the older B model Iroquois were replaced by higher capacity D and H models. Additional ground staff were posted in from Australia and, for the first time, Royal New Zealand Air Force and Royal Australian Navy pilots joined the unit.

To provide fire support to Australian ground forces, 9 Squadron received approval to convert four of its Iroquois into gunships – heavily armed attack helicopters – during March 1969. Despite the acquisition of gunships, most of 9 Squadrons more eventful missions revolved around the "hot" extraction of troops under fire and Dust Offs where casualties had to be winched out of the jungle while the helicopters were often under fire. On several occasions helicopters were either shot down or badly damaged while engaged in these hazardous operations.

Such a mission occurred on February 13 1970 when Flying Officer M Haxell and crew were tasked to extract a patrol which could not break contact with the Viet Cong. Arriving over the landing zone, the helicopter came under such intense fire that the patrol commander advised Flying Officer Haxell to vacate the area and that the troops would "try to make a run for it". Scorning such a course of action, Haxell put the helicopter down into the landing zone while his gunners suppressed the enemy's fire. The soldiers, still firing their weapons, leapt aboard the Iroquois and, as the helicopter took off, several enemy soldiers were killed in the landing zone.

Another loss occurred on April 17 1971 when Flying Officer M Castle and his crew were shot down while trying to evacuate a wounded South Vietnamese soldier in the notorious Long Hai

Combined operations – a 9 Squadron Iroquois hovers over Australian Centurion tanks and Armoured Personnel Carriers in Vietnam.

hills. An army medical orderly, part of the helicopter's crew, was killed as were two soldiers on the ground who were hit by flying debris. The downed crew, one of whom was injured, were later evacuated by another Australian helicopter.

9 Squadron flew its find Vietnam combat mission on November 19 1971 and sailed for Australia on December 8. The unit settled into its new base at Amberley and by early 1972 was operating in the varied roles typical of a helicopter squadron. Army support tasks predominated but 9 Squadron again flew extensively conducting search and rescue tasks for the civilian community.

Many exercises were carried out and the Iroquois were, at times, deployed outside Australia – principally to New Guinea and other South West Pacific island nations. On several occasions 9 Squadron supported police drug operations by airlifting police into inaccessible bushland where drug plantations had been established.

In early 1982, 9 Squadrons operation's were disrupted when eight helicopters and a contingent of personnel were deployed to the Middle East on peacekeeping duties as part of the Multi National Force and Observers – a commitment maintained until 1986.

During February 1988, 9 Squadron began to re-equip with Sikorsky S-70A Black Hawk helicopters but after conversion training had been completed the unit moved to Townsville where it disbanded on 14 February 1989. The unit's personnel and helicopters were then used to form the nucleus of the Army's 5 Aviation Regiment.

Formed: *January 1 1939*
Squadron Code: *I, YQ*
Aircraft: *Seagull, Walrus, Iroquois, Black Hawk*
Locations: *Richmond: Jan 1939 – Jan 1940*
Rathmines: January 1940 – December 1944
Disbanded:
Williamtown: June – November 1962
Fairbairn: November 1962 – June 1966
Vung Tau: June 1966 – December 1971
Amberley: December 1971 – December 1988
Townsville: December 1988 – February 1989
Disbanded: *February 14 1989*

---

10 Squadron formed on July 1 1939 at Point Cook. The new general reconnaissance unit, commanded by Squadron Leader L Lachal, was temporally equipped with Supermarine Seagull V amphibians, a de Havilland DH.60 Moth floatplane and a Supermarine Southampton flying boat pending the delivery of Shorts Sunderland flying boats.

Before the end of July drafts of pilots and ground staff departed for the United Kingdom to accept the Sunderlands and gain experience on the type before ferrying them back to Australia. On completing their training, the Australians assembled at Pembroke Dock, where, after war was declared, they were to remain on active service with RAF Coastal Command. The decision not to return 10 Squadron to Australia resulted in it becoming the first Dominion squadron to go into action in World War II.

10 Squadron was judged operational on February 1 1940 and immediately began operations. The unit's main tasks comprised convoy escorts, anti-submarine patrols, air sea rescue work and ferrying equipment and personnel on long range transport missions. Prior to the arrival of the additional airmen despatched from Australia, 10 Squadron was still short of personnel of all categories and ground staff personnel regularly flew on these missions in virtually all crew categories. The copilot shortage was so acute that even the Equipment Officer – who had no flying experience at all – was pressed into service in this capacity!

On April 1 1940, 10 Squadron moved to Mount Batten where, on May 14, Flight Lieutenant H Birch's crew located the unit's first U Boat. In June the first attack against a U Boat occurred when Flight Lieutenant C Pearce's crew superficially damaged a submarine. Next month, on July 1 Flight Lieutenant G Havyatt's crew gained the distinction of sinking 10 Squadron's first submarine when they destroyed the German Boat U-26. This sinking was only the second by a Coastal Command aircraft since the outbreak of hostilities.

July also saw the commencement of a major and long term detachment to Oban, Scotland, which enabled the Sunderlands to cover shipping movements to the north of the UK.

Despite some damaging air attacks on Mount Batten which destroyed aircraft, hangars and facilities, 10 Squadron's ground staff maintained an extremely high serviceability rate which similar RAF squadrons could not match. Their efforts were directly responsible for the unit flying a tremendous number of flying hours during the dark days of 1940. After France surrendered, a heavy anti-invasion patrol commitment had to be maintained and it became common for the flying boats to be attacked on their patrols. During one mission on March 5 1941, 10 Squadron claimed its first aerial victory when one of two attacking Junkers Ju 88s were shot down. In another successful combat on June 5, Flight Lieutenant G Thurston's gunners destroyed two Arado Ar 196 floatplane fighters. Through incidents such as these the Sunderlands became known to the Germans as "flying porcupines".

June also saw the allocation of three former BOAC G Class flying boats which were intended for long distance transport tasks. Unfortunately when converted for military service these aircraft had poor performance and were withdrawn after a short time.

Patrols continued at an intensive pace but submarines were rarely sighted, although the wreckage from ships and survivors in lifeboats were frequently found and, where possible, ships directed to their rescue. At times downed aircrews were also located and occasionally the Sunderlands were able to land and pick up these airmen if the seastate was not too rough. Such open sea landings required a mix of skill and good luck as the flying boats were only designed for operation from sheltered waterways.

On June 21 1942 Flight Lieutenant M Judell's Sunderland was shot down by Arado fighters whilst engaged on an anti-submarine patrol over the Bay of Biscay. This was the first 10 Squadron Sunderland lost to enemy action since the beginning of the war. Mid 1942 also saw a rash of submarine sightings resulting in a number of attacks in which several submarines were damaged.

Operations continued into 1943 with occasional attacks against U Boats and regular combat with German fighters, which usually operated in large formations. On May 7 Flight Lieutenant G Rossiter's crew sighted three submarines in a single patrol and executed a successful depth charge attack against the third U Boat (U-465) which later sank. On May 17 Flight Lieutenant K McKenzie's crew failed to return from a patrol, presumed shot down by fighters. Later that day, Flight Lieutenant J Weatherlake's Sunderland returned to base with 200 bullet holes after being attacked by Ju 88s.

In an attempt to counter the effectiveness of these fighters, 10 Squadron developed a number of armament modifications. One of these was the

One of 10 Squadrons "Flying Porcupines" on patrol.

introduction of 50 calibre galley hatch mounted machine guns. Another involved the fitment of four .303 machine guns in the Sunderlands wings for use against submarines with the dual purpose of suppressing anti-aircraft gun crews and also to aid in the reduction of line errors when making depth charge attacks.

A different kind of modification destined to have wide ramifications in the RAF was the re-design of the Sunderland's engine mounts and wing internal layout to enable the aircraft's un-reliable Bristol Pegasus engines to be replaced by Pratt & Whitney powerplants. The plans drawn up by 10 Squadron's ground staff were accepted almost in toto by Short Brothers – the Sunderland's manufacturers – and led to the pro-duction of Pratt & Whitney engined Sunderlands which entered service both with the RAF and RAAF towards the end of the war.

A "long nosed" Lincoln maritime patrol aircraft at its Townsville base during the 1950s.

On May 31 1943 Flight Lieutenant M Main-prize's crew sank U-563 in conjunction with three aircraft from other squadrons. A remarkably determined attacked occurred on July 9 when Flying Officer D Grey's Sunderland attacked three U Boats sailing in company for stronger anti-aircraft protection. For good measure the U Boats also had an escort of three Ju 88s! Al-though their attack was unsuccessful the crew displayed extraordinary courage and determina-tion in attacking so well a defended target.

On August 1 an attack by Flight Lieutenant R Fry's crew resulted in the destruction of yet another U Boat (U-454). Unfortunately the sub-marines gunners also found their mark and the Sunderland, badly hit on its attack run, crashed into the sea immediately after dropping its depth charges. Six of the flying boats 12 man crew, in-cluding its gallant captain, died in the crash.

The latter part of 1943 saw a continuance of Ju 88 formation attacks against 10 Squadron air-craft. On August 3 Flying Officer A Williams' crew was engaged by seven Ju 88s. In the run-

A formation of 10 Squadron Neptunes.

ning fight, some of the enemy fighters were hit but the Sunderland's front gunner was killed and three other crewmen wounded. On August 8, Flight Lieutenant N Gerrard's crew survived a similar encounter only to disappear three days later while on another patrol. Several crews were lost during this period in similar circumstances, the radio operator of one managing to send part of a distress call before his aircraft was shot down.

On November 30 Flight Lieutenant T Clarke's Sunderland was attacked by six Ju 88s. Three crew members were wounded in the action and the flying boat was so severely damaged that it had to glide in for a landing after its fuel tanks were punctured by gunfire. The Sunderland lived up to its reputation as the "flying porcupine" on this occasion by leaving one Ju 88 with both engines burning and a second with one engine on fire. These two aircraft had very little chance of returning to their French airfields.

In December 1943 10 Squadron expended considerable effort searching for blockade running ships. While few sightings resulted from these missions, on one occasion a Sunderland captained by Flight Lieutenant J McCulloch located several German destroyers and alerted a British cruiser force which hurried to the area and sank three of the German warships.

The new year began very successfully for 10 Squadron when, on January 8, Flying Officer J Robert's crew located an enemy submarine on the surface. Attacking through heavy anti-aircraft fire, the Sunderland's wing mounted machine guns in conjunction with the front turret gunner swept the enemy gun crew from the deck as the aircraft executed an extremely accurate attack which ultimately sank the submarine.

During February 10 Squadron created a Coastal Command record by flying 1143 hours, a remarkable rate of effort and one which only could be sustained by the dedicated efforts of the ground staff. Flight Lieutenant McCulloch's crew created another record – at least for 10 Squadron – when they were attacked by 16 Ju 88s on February 15. After damaging some of their opponents, the flying boat escaped into cloud, but not before its tail gunner had been killed.

On July 8, Flying Officer W Tilley's crew attacked and badly damaged a U Boat. They then homed other aircraft to the scene and further attacks by a Sunderland captained by Flight Lieutenant R Cargeeg and an American Liberator completed the German submarine's destruction.

Anti-submarine patrols and patrols in support of invasion shipping continued throughout 1944 but the new "Schnorkeling" equipment fitted to German submarines had from this time no effective countermeasure and very few sightings were made for the rest of the war. As the Allied bridgehead in France expanded, the Germans were forced to abandon the airfields from which they had mounted such heavy attacks on Coastal Command aircraft. Consequently, the frequency of combat rapidly decreased and by late 1944 were almost non existent.

10 Squadron flew its last wartime mission on May 7 1945 and the unit disbanded on October 26 1945. 10 Squadron had sunk six submarines

and was the only RAAF unit on continuous active service for the duration of World War II in Europe.

While 10 Squadron was still at Mount Batten, a second 10 Squadron came into existence at Driffield on June 20 1945 when 466 Squadron was renumbered 10 Squadron. The unit had ceased to be part of Bomber Command and its Halifax IVs and Liberator VIIIs were to be used in the transport role. These plans came to naught however, and the unit disbanded at Bassingbourn on October 26 1945,

A P-3C Orion in 10 Squadron markings (David Foote)

the same day as the "real" 10 Squadron (which officially had become a detachment of the Halifax equipped unit).

10 Squadron reformed at Townsville on March 1 1949. The unit, as in World War II, was to operate in the maritime patrol and anti-submarine role. Initially, equipped with Australian made Avro Lincoln B.30s – the standard bomber version – 10 Squadron later received Mk.31 aircraft and finally specially modified MR.31s (the so called "long nosed" Lincoln) during 1953. With these aircraft 10 Squadron patrolled Australia's northern waters and much of the Pacific Ocean – a massive area representing about one tenth of the world's surface.

The Lincolns also supported the civil community in various emergencies and proved highly suited to search and rescue tasks. During 1961 serious corrosion was found in the aircraft's wings – a legacy of many 12 to 15 hour overwater flights – and the aircraft had to be prematurely grounded. Although it had been announced that 10 Squadron was to be re-equipped with Lockheed P2V-7 Neptunes in October 1959, the new aircraft had not yet arrived and it was not until March 1962 that the first were received.

With its Neptunes, 10 Squadron continued its anti-submarine, maritime reconnaissance and search and rescue roles. The Neptunes regularly participated in defence exercises with American, British and New Zealand forces and, during 1965, were used to provide anti-submarine escort to HMAS Sydney when that ship carried Australian troops to Vietnam.

By the 1970s the Neptunes were beginning to show their age and orders were placed for Lockheed P-3C Orions as replacement aircraft. From July 1977, 10 Squadron began to wind down its operations prior to receiving its new aircraft. In

early 1978, 10 Squadron moved to Edinburgh where the Orions were delivered between May 1978 and January 1979.

10 Squadron continues to operate in its traditional roles and, since being fitted with the potent Harpoon missile in 1981, anti-shipping has been added to the unit's tasks. Regular detachments to Butterworth, Malaysia are a feature of 10 Squadron's operations and today, along with the Orions of 11 Squadron, the unit provides Australia with its general airborne maritime patrol and anti-submarine warfare capability.

Formed: *July 1 1939*
Squadron Code: *K, RB*
Aircraft: *Seagull, Gipsy Moth, Southampton, Sunderland, G Class, Halifax, Liberator, Lincoln, Neptune, Orion*
Locations: *Point Cook: July 1939 – Dec 1939*
Pembroke Dock: December 1939 – April 1940
Mount Batten: April 1940 – June 1941
Pembroke Dock: June – December 1941
Mount Batten: December 1941 – June 1945
Disbanded:
Driffield: June – September 1945
Bassingbourn: September – October 1945
Disbanded:
Townsville: March 1949 – January 1978
Edinburgh: January 1978 – present

---

11 SQUADRON

Formed at Port Moresby on March 21 1939 under the command of Flight Lieutenant J Alexander, 11 Squadron was initially equipped with four aircraft – two Short Empire S.23 flying boats and two Supermarine Seagull V amphibians. With this handful of aircraft, the new general reconnaissance squadron was to conduct long range reconnaissance patrols to the north of Australia with particular emphasis on observing Japanese shipping movements in the region.

One early task undertaken was an unsuccessful search for the Italian merchant ship *Romolo* which had sailed from Brisbane within hours of Italy's declaration of war. Up until Japan's entry in the war during December 1941, 11 Squadron's activities consisted of routine tasks, comprising reconnaissance patrols and transport flights.

In March 1941, it began to re-equip with Consolidated Catalina flying boats, but the six delivered before August were taken over by 20 Squadron (11 Squadron being left with Empires) which formed at Port Moresby during this period. By Christmas, however, 11 Squadron was again in the process of re-equipping with Catalinas.

In early November 1941 a number of flying boats were detached to Western Australia where they conducted fruitless searches for the missing cruiser HMAS *Sydney*. It was later found that *Sydney* had been lost with all hands in an engagement with the German surface raider *Kormoran*.

Within hours of Japan's entry in the war, 11 Squadron sustained the RAAF's first casualties of the Pacific campaign when Flight Lieutenant L Sloan's Catalina crashed into a hill during a night takeoff from Port Moresby on December 8. There were no survivors.

The immediate success of Japanese offensive action ensured that 11 Squadron flew at a very intensive rate indeed. Operating out of advance operating bases, long range patrols of up to 20 hours duration were a feature of the unit's work, as were night bombing attacks on Japanese island strongholds. Notwithstanding enemy opposition and the loss of some aircraft, 100 sorties were flown in December and 115 in January 1942.

As the Japanese maintained their southward thrust, the Catalinas and Empires evacuated military personnel and civilians caught in the path of the advancing enemy. By February 1942 Port Moresby itself came under attack and, while personnel losses were light, the destruction of several flying boats on the water represented a major loss to the unit, and indeed the RAAF which then possessed very few long range patrol aircraft.

Another important task allocated to 11 Squadron at this time was the provision of experienced pilots to fly in the lead aircraft of newly arrived American bomber squadrons during their early attacks on Japanese bases. The Australians were thus available to assist the inexperienced Americans in locating their targets and to provide advice on combat tactics. In early 1942, 11 Squadron also conducted the first open sea rescue of the Pacific war when Flight Lieutenant T Duigan's crew alighted and recovered a downed American bomber crew.

By now, Port Moresby was under such heavy attack that 11 Squadron was withdrawn to Bowen where it continued to operate intensively. Targets attacked included Lae, Guadalcanal, Gasmata, Madang, Tulagi, Gona, Salamaua and Buna. Reconnaissance and anti-submarine patrols also continued and during one of these, on January 6 1943, Flight Lieutenant D Vernon's

11 Squadron Catalinas. The closest flying boat (without blisters) is a former Dutch aircraft flown to Australia to avoid capture by the Japanese.

crew bombed and sank a 5100 tonne Japanese vessel.

On the night of March 1, Catalinas staging through Milne Bay shadowed a large Japanese convoy during the Battle of the Bismarck Sea. Next day, the convoy was almost completely destroyed in one of the decisive actions of the South West Pacific campaign.

April 22 saw a mixed formation of 11 and 20 Squadron Catalinas carry out the RAAF's first mine laying operation when magnetic mines were successfully laid near Kaivieng. This mission marked the commencement of a highly successful mining campaign which was responsible for the sinking of ships, the disruption of maritime trade and the closure of ports. So successful proved the campaign that it was intensified and maintained until the end of the war.

11 Squadron moved to Rathmines in July 1944 where it was employed on anti-submarine patrols off Australia's east coast. The Catalinas, however, remained available for mine laying operations and in this capacity 11 Squadron despatched six Catalinas to participate in the RAAF's largest mine laying operation of World War II. Staging through San Pedro Bay with 19 other flying boats from 20, 42 and 43 Squadrons, the Catalinas laid over 50 mines in Manila Bay. Coincidentally, with a round trip of some 14,500 kilometres, the 11 Squadron aircraft also flew the RAAF's longest operation of World War II.

Operational flying continued up until the Japanese surrender, after which flying commitments gradually decreased and, after earning a distinguished war record, 11 Squadron disbanded at Rathmines on January 15 1946.

11 Squadron operated Australian built Lincolns for a short period prior to converting to Lockheed Neptunes.

In over twenty years of Orion operations, 11 Squadron has only lost one Orion and that in a ground accident not in the air. The drama of that event is seen at Edinburgh in January 1984. (RAAF Museum)

On July 1 1948, the Catalina equipped Search and Rescue Wing at Rathmines, was renamed 11 Squadron. From the beginning flying was fairly limited, however, in early August Flight Lieutenant B Delahunty's crew conducted a significant mission when they operated from Macquarie Island in Antarctica, ferrying personnel from Australia in a Catalina fitted with jet assisted takeoff modifications.

By April 1950, 11 Squadron was no longer operational and was reactivated at Amberley in November. Re-equipped with Avro Lincoln Mk.30s, it almost immediately deployed to Pearce to conduct maritime patrols over the Indian Ocean. The Lincolns were only interim equipment, however, and 11 Squadron received its first Lockheed P2V-4 (followed by P2V-5) Neptunes in November 1951. Deliveries of the new aircraft were not completed until January 1953 and during this period flying was considerably hampered with large drafts of personnel in the United States on Neptune conversion courses.

11 Squadron's sojourn in Western Australia was a brief one, ending in May 1954 when the unit moved to Richmond. From here exercises, some of which involved overseas deployments, were a regular feature of its work. Another important task was the provision of aircraft during searches for missing civilian vessels. One unusual task undertaken was the provision of aircraft to conduct security patrols in support of the British atom bomb tests at Monte Bello Island in May/June 1956.

Later, in February 1957, three Neptunes participated in "Operation Westbound" – the RAAF's first around the world flight. Between November 1958 and January 1959, 11 Squadron also provided search and rescue support for two squadrons of RAAF Sabres deploying to Butterworth.

Routine anti-submarine exercises, maritime patrol work and search and rescue tasks kept the unit busy into the 1960s. By mid 1967, 11 Squadron was preparing for its conversion from Neptunes to Lockheed P-3B Orions. During this transitional period, while many personnel were again in the USA on conversion and familiarisation courses, the unit moved to Edinburgh where the first Orions arrived in early 1968.

The B model Orions provided sterling service for some 12 years until replaced by P-3C Orions in 1985/86. Operating these aircraft today, 11 Squadron continues in its anti-submarine warfare/maritime patrol role and armed with the lethal Harpoon missile provides Australia with an invaluable long range anti-shipping capability.

Formed: *March 21 1939*
Squadron Code: *L, FJ*
Aircraft: *Empire, Seagull, Catalina, Lincoln, Neptune, Orion*
Locations: *Port Moresby: Mar 1939 – May 1942*
Bowen: May 1942 – November 1942
Cairns: November 1942 – July 1944
Rathmines: July 1944 – January 1946
Disbanded:
Rathmines: July 1948 – April 1950
Amberley: April – November 1950
Pearce: November 1950 – May 1954
Richmond: May 1954 – January 1968
Edinburgh: January 1968 – present

---

12 SQUADRON

On February 6 1939, 12 Squadron under the command of Squadron Leader C Eaton, a combat experienced veteran of World War I, formed at Laverton. Equipped with Avro Anson Mk.Is and Hawker Demon Mk.Is the unit began moving to Darwin in July 1939 and became the first RAAF squadron to be permanently based in the Northern Territory.

By late August seven Ansons had arrived at Darwin's civil aerodrome and the first reconnaissance patrol was flown two days later. Meanwhile at Laverton, the Demons had been replaced by Australian built CAC Wirraways and on September 1 these aircraft departed for Darwin arriving four days later.

Effective from June 1 1940, 12 Squadron underwent a major reorganisation when it was directed to form RAAF Station Darwin and 13 Squadron from within unit resources. Almost all of the Squadron Headquarters personnel were used to form Station Headquarters while the Ansons of A and B Flight became the nucleus of the newly constituted 13 Squadron. Only the Wirraways of C Flight remained as 12 Squadron.

In April 1941, 12 Squadron, built up to full strength, and reassigned a general purpose role, moved to Darwin's newly constructed RAAF airfield. Operations from here were the same as previously carried out including anti-submarine patrols, shipping escorts, coastal patrols – which covered an area from Port Hedland to Millingimbi – and seaward reconnaissance.

For longer distance flights, refuelling facilities were maintained at Port Hedland, Wyndham, Derby, Broome, Bathurst Island, Millingimbi and Drysdale River Mission. The Wirraways were also engaged in the surveillance of Japanese pearling luggers which were suspected of gath-

Having operated Wirraways for a considerable time, 12 Squadron later re-equipped with Vultee Vengeance Dive bombers.

ering information for military purposes as war in the Pacific became more probable.

After war with Japan commenced, 12 Squadron intensified its flying effort. B and C Flights were dispersed to Batchelor while A Flight returned to the civil airfield. This was the unit's disposition when, on February 19 1942, the Japanese launched two massive air attacks against Darwin. 12 Squadron lost two Wirraways on the ground and a considerable quantity of technical equipment and stores when its new hangar was gutted by fire.

Following these raids, 12 Squadron was kept busy dropping supplies to stranded survivors from sunken ships. Anti-submarine patrols outside of the harbour were also flown and, as the threat of invasion appeared imminent, army co-operation tasks were practised to an increasing extent. Despite further air raids, 12 Squadron continued its tasks without encountering any direct opposition from the Japanese – a situation which, as the months went by, began to aggravate the Australian airmen.

In July 1942, 12 Squadron moved to Pell, south of Darwin but, within two months, was again at Batchelor where the following month word was received that the unit would re-equip with Vultee Vengeance dive bombers. This decision was a popular one as many now thought that they would at least be able to take an active part in the war. 12 Squadron was the first unit in the South West Pacific to receive the Vengeance and, later was the first to use them in combat. Unfortunately, even after receiving its full allocation

of dive bombers the unit was still not employed on active operations, routine patrols and searches for downed Spitfire pilots taking up most of the Vengeance crew's time.

On June 18 1943, excitement was intense when 12 Squadron finally despatched 12 aircraft on its first – and as it turned out last – dive bombing mission. Photo reconnaissance had detected the Japanese building an airfield on Selaru Island only 500km from Darwin and the Vengeances were tasked to destroy the villages in which the airfield construction workers were housed. The strike was accurately made and, although one Japanese fighter was sighted, it did not attack.

In late 1943, 12 Squadron moved to Marauke on Dutch New Guinea where monotonous but essential anti-submarine patrols and shipping escorts were the unit's lot. The only exception to this galling state of affairs occurred on October 9 1943 when a Vengeance sighted a Jake floatplane and exchanged machine gun fire with it at long range.

In July 1944, 12 Squadron began moving to Strathpine where it was reduced to cadre pending re-equipment with Consolidated Liberator heavy bombers and, after a further move to Cecil Plains, rearming commenced. By the end of February 1945, 12 Squadron had 80 officers, 506 airmen and 10 Liberators on strength (compared with just one officer four airmen and no aircraft just two months previously).

By late March 1945, 12 Squadron was again on the move to Darwin, a task completed early in May. Operations began on May 24 and the unit quickly increased its operational effort conducting strikes against barges and shipping around Timor and in the Banda and Arafura Seas.

With the Japanese surrender in August, 12 Squadron was busily engaged in dropping supplies of medicine and food to Allied POWs throughout the Dutch East Indies. Later the Liberators were used to repatriate some of these unfortunate individuals to Australia.

By March 1946, 12 Squadron's role in the islands was completed and the unit moved to Amberley where it later re-equipped with Australian built Avro Lincolns before being renumbered 1 Squadron on February 23 1948.

12 Squadron reformed at Amberley on September 3 1973. The new 12 Squadron was to principally be an army support unit equipped with twin rotor Boeing Vertol CH-47C Chinook medium lift helicopters. However, the Chinooks were not delivered until March 1974 and the first flight did not occur until July 8.

The versatility of the Chinooks ensured that their tasks were varied to say the least. Usually, they operated in Northern Australia, lifting artillery pieces and carrying troops, however, civil aid activities also accounted for a great deal of the Chinooks' flying effort. Tasks included air-sea rescues, placing lighthouses on isolated parts of the coastline, delivering fodder to cattle stranded by flood water and placing power poles and air conditioner plants on the tops of skyscrapers. On two occasions 12 Squadron supported the Queensland Police by carrying seized narcotics and delivering fuel supplies at an inland location to refuel RAAF Iroquois helicopters which deployed the Police into remote drug plantation areas.

On August 25 1980, a Chinook captained by Wing Commander J Dahlitz departed Amberley on what to date remains the Australian Defence Force's longest helicopter operation. The helicopter flew to a site 100km east of Butterworth, Malaysia where a crashed Malaysian Nuri S-61 helicopter was recovered, flown to Ipoh and later to Kuala Lumpur for major repairs.

Due to a need to reduce defence expenditure and with the introduction of the Black Hawk helicopter (which it was hoped would fulfil some of the Chinook's roles) the decision was made to cease flying as at June 30 1989 and to disband the unit. Subsequently 12 Squadron officially disbanded on August 25 1989.

Formed: February 6 1939
Squadron Code: M, NH
Aircraft: Demon, Anson, Wirraway, Vengeance, Liberator, Lincoln, Chinook
Locations: Laverton: February – Sept 1939
Darwin: September 1939 – February 1942
Batchelor: February – July 1942
Pell: July – September 1942
Batchelor: September 1942 – October 1943
Marauke: October 1943 – July 1944
Strathpine: July – December 1944
Cecil Plains: December 1944 – May 1945
Darwin: May 1945 – March 1946
Amberley: March 1946 – February 1948
Disbanded:
Amberley: September 1973 – August 1989
Disbanded: August 25 1989

The Chinooks of 12 Squadron were the largest and most versatile helicopters ever operated by the RAAF. (RAAF Museum)

13 SQUADRON

On June 1 1940, 13 Squadron at Darwin formed from two Anson Mk.I flights and personnel detached from 12 Squadron.

The new unit under the command of Flight Lieutenant J R Balmer was allocated a general reconnaissance role and undertook shipping patrols and searches as well as security patrols across northern Australia.

In June 1940, the first Hudson Mk.I and Mk.IIs arrived and the unit continued its patrol work with little incident. May saw the commencement of familiarisation flights over the Netherlands East Indies, where it was anticipated that 13 Squadron would operate in the event of hostilities commencing with Japan. In November 1941 three Hudsons engaged in fruitless searches for the Australian Navy Cruiser HMAS *Sydney* which had been sunk by the German shipping raider *Kormoran*.

The expected commencement of hostilities with Japan saw 13 Squadron divided, with a detached flight of six Hudsons already at Laha on Ambon and another six about to depart from Darwin for Namlea. Patrols were intensified by these East Indies detachments.

On December 23 patrolling Hudsons located a Japanese convoy at the northern tip of the Celebes. This was a particularly important sighting and resulted in a number of attacks being mounted on the Japanese vessels.

13 Squadron's luckiest man must surely have been Sergeant B Hack. On January 1 1942 the Hudson in which he was a crewmember suffered an engine failure while on a reconnaissance patrol. At low level, the aircraft's captain, Flying Officer J Turnbull, jettisoned the bomb load which unfortunately detonated and blew up the aircraft. All the crew were killed except Sergeant Hack who was catapulted through the fragments of the Hudson and into the sea. An accompanying Hudson, observed the injured airman in the water and after 15 minutes he was dragged into a Catalina and returned to Ambon. Ensconced in the base hospital, he endured many Japanese air raids, and doctors, assessing the increasingly dangerous situation, evacuated him to Darwin – just in time for the first massive attack on that base!

By January 1942, Laha and Namlea were under frequent attack, firstly by Japanese flying boats and later by land based aircraft. Despite an almost total lack of fighter and anti-aircraft defences, operations continued with some success and, on January 10, Wing Commander J Ryland's crew intercepted one of the raiding flying boats and gamely attacked the heavily armed aircraft, scoring numerous hits, until forced to break off the action due to fuel shortage.

Just two days later 13 Squadron suffered a heavy loss when two Hudsons (captained by Flight Lieutenants G Sattler and A R Barton) failed to return from a strike against Japanese shipping at Menado in the Celebes. These casualties were only the first in a succession sustained by the unit which was by now operating under extremely difficult conditions due to almost daily attacks on its forward bases.

The first weeks of February saw the surviving Hudsons return to Darwin where they continued

13 Squadron was the only RAAF unit in the Pacific to operate Venturas.

50

their vital reconnaissance flights and bombing missions. To maintain this operational effort the ground staff toiled constantly to keep the aircraft airworthy despite chaotic conditions, lack of spare parts and almost nonexistent maintenance facilities. The first Japanese air raid on Darwin on February 19 had a major impact on operations as the unit's headquarters, stores and spares were destroyed. Following this attack, a partial withdrawal to Daly Waters was effected.

Air raids at Darwin continued and some of these caused casualties to the portion of 13 Squadron which remained there. Aircraft numbers were now so low that until replacements were received only one or two were available each day. For its part in these critical operations against overwhelming odds, 13 Squadron was later awarded the United States Presidential Unit Citation – an honour bestowed on only two RAAF units – 13 and 2 Squadrons.

In April 13 Squadron began to attack targets in Japanese occupied Timor and the East Indies while continuing photographic reconnaissance and search flights. These missions were interrupted when the unit moved to Hughes strip in the first days of May.

On August 10 the Hudsons conducted a most successful shipping strike off the South Coast of Timor. Flight Lieutenant Trewin's crew scored direct hits on a 1220 tonne ship after which they attacked and definitely sank a second vessel. Squadron Leader Moran's crew were also successful sinking a 3060 tonne ship.

By early 1943, 13 Squadron was urgently in need of rest and re-equipment. Accordingly, it became non-operational in April and, after handing over its aircraft to other units, moved to Fairbairn where it was re-equipped with Bristol Beauforts and Lockheed Venturas.

For a few weeks in November and December a detached flight of Beauforts, carrying out anti-submarine patrols, was maintained at Coffs Harbour while another flight, comprising three Venturas, conducted similar duties from Camden.

In May/June 1944 13 Squadron, now fully equipped with B-34 and PV-1 Venturas, moved to Cooktown but only weeks later again moved, this time to Gove in the Northern Territory. From here the Venturas carried out anti-submarine patrols and conducted strikes around Timor and the Netherland East Indies.

From Gove, 13 Squadron moved to Moratai and hence to Labuan. After Japan's surrender the Venturas evacuated POWs from enemy held areas as well as returning other personnel to Australia. Courier flights both to Singapore and to Darwin were also undertaken before 13 Squadron disbanded on January 11 1946.

As a non flying RAAF Reserve unit, 13 Squadron reformed at Darwin on July 1 1989. In June 1990, the unit was finally presented with the Presidential Unit Citation which it had been awarded 48 years previously.

Formed: *June 1 1940*
Squadron Code: *N, SF*
Aircraft: *Anson, Hudson, Beaufort, Ventura*
Locations: *Darwin: June 1940 – Feb 1942*
Daly Waters: February – May 1942
Hughes: May 1942 – April 1943
Fairbairn: April 1943 – July 1944
Cooktown: July – August 1944
Gove: August 1944 – August 1945
Labuan: August 1945 – January 1946
Disbanded:
Darwin: July 1989 – present

---

14 SQUADRON

On February 6 1939, 14 Squadron formed at Pearce. Initially equipped with Avro Anson Mk.I aircraft, the unit, commanded by Flight Lieutenant C W Pearce, spent the lead up to World War II engaged in navigational night flying, army co-operation and other training.

After the outbreak of hostilities 14 Squadron conducted seaward reconnaissance flights and from early 1940 anti-submarine patrols over convoys carrying Australian troops to the Middle East. To make this task easier the unit re-equipped with Hudson Mk.IVs in mid 1940.

14 Squadron continued its routine duties – at times maintaining a detachment at Albany – and

staging through other airfields until after the Japanese entered the war in December 1941. Thereafter, seaward and anti-submarine patrols were intensified but generally produced little result. On one occasion, however, a submarine, which later proved to be American, was attacked and severely damaged (a court of inquiry later exonerated the pilot of blame). In two other incidents, enemy aircraft were encountered but in each case the Hudsons were unable to close with them.

On March 3 1942, 14 Squadron had its only real brush with the enemy when a Hudson was destroyed on the ground at Broome by Japanese

An early Avro Anson. Despite its short range, 14 Squadron operated the type on maritime reconnaissance flights.

fighters which made a surprise attack on that town. Although the crew escaped injury, casualties and damage in the raid were heavy.

In late 1942, 14 Squadron rearmed with Australian built Beauforts, the first patrols with the new aircraft being flown in December. Ranging far out to sea the Beauforts, as had the Hudsons before them, covered a massive area of ocean in their uneventful flights.

During March 1943 reconnaissance patrols were considerably strengthened due to an expected Japanese naval foray – which did not eventuate – into the Indian Ocean. From February 1944 another scare, involving a German submarine supposedly operating from the Dutch East Indies, for a time required an increased rate of effort but no sightings were made. 14 Squadron continued its important, if unspectacular, work until the cessation of hostilities and disbanded at Pearce on October 30 1945.

Formed: February 6 1939
Squadron Code: P, PN
Aircraft: Anson, Hudson, Beaufort
Locations: Pearce: Feb 1939 – October 1945
Disbanded: October 30 1945

15 SQUADRON

Equipped with Beauforts and under the command of Wing Commander I L Campbell, 15 Squadron formed at Camden on January 27 1944. The unit operated in the anti-submarine and convoy escort role off Australia's east coast.

Operations remained routine almost until the end of the war, however, in April 1945, most of the unit deployed to Tadji joining other Beaufort Squadrons attacking targets around Wewak. Operating at the end of a tenuous supply line which forced the detachment at times to utilise captured Japanese bombs, the Beauforts undertook a series of very accurate strikes against Japanese positions. To support this detachment's operations, 15 Squadron headquarters moved to Madang.

While active operations were being undertaken by the New Guinea detachment, other detachments were operating from Cairns and Townsville, while a rear party remained at Camden.

In October 1945, 15 Squadron moved to Kingaroy where it disbanded on March 23 1946.

Formed: January 27 1944
Squadron Code: DD
Aircraft: Beaufort
Locations: Camden: January – February 1944
Menangle: February – May 1944
Camden: May 1944 – April 1945
Madang: April – November 1945
Kingaroy: October 1945 – March 1946
Disbanded: March 23 1946

15 Squadron operated Beauforts during its brief history.

18 SQUADRON

One of the RAAF's most unusual units, 18 Squadron formed at Fairbairn on April 4 1942. Commanded by Major B J Fiedeldij, the new RAAF squadron contained a mix of Netherlands East Indies personnel (both Dutch nationals and Javanese) and Australians.

Allocated a bomber role, 18 Squadron trained on Douglas Boston (a mixture of DB-7Bs and A-20As) and North American B-25C Mitchell aircraft before standardising on the latter. Although aircraft captains were always Dutch, many of the aircrews and over half the ground staff were RAAF members who were grouped under their own Commanding Officer for disciplinary and administrative purposes.

During its conversion training, anti-submarine patrols were conducted on a regular basis. During one of these a Mitchell attacked and, it was thought – incorrectly – sank a Japanese submarine on June 5 1942.

On July 6 1942, 18 Squadron was officially deleted as a unit of the RAAF and became part of the Netherlands East Indies Forces. The unit moved to the Darwin area in January 1943 and, under the operational control of RAAF Command, participated in many attacks and anti-shipping strikes, especially over the Netherlands East Indies.

After the war, RAAF personnel were withdrawn from 18 Squadron which returned to the East Indies. In 1948 the Dutch unit was again in action, this time against Indonesian Nationalists. When Indonesia finally won its independence from the Dutch, 18 Squadron disbanded on June 25 1950 and its aircraft and personnel were absorbed into the Indonesian Air Force.

Formed: *April 4 1942*
Squadron Code: *GM*
Aircraft: *Boston, Mitchell*
Locations: *Fairbairn: April – January 1943*
Deleted as a RAAF Squadron: *July 6 1942*

---

20 SQUADRON

20 Squadron formed at Port Moresby under the command of Squadron Leader W N Gibson on August 1 1941. Initially equipped with Short Empire flying boats, the new unit, in close co-operation with 11 Squadron – also at Port Moresby – operated in the general reconnaissance role. War with Japan was expected anytime and the Empire Flying Boats, which before the end of the year were replaced by Consolidated Catalinas, flew long range patrols searching for and reporting shipping movements.

After the outbreak of hostilities, patrols were intensified and bombing raids against Japanese occupied territory commenced. The Catalinas were not especially suited to this role but they were the only RAAF aircraft possessing the range required to strike at some of these distant targets. Another important task undertaken was the evacuation of civilians from Port Moresby, Rabaul, Salamaua and other locations in New Guinea.

An early loss occurred on January 21 1942 when Flight Lieutenant R Thompson's crew was shot down by four Zero fighters while searching for a Japanese aircraft carrier task force. The flying boat located the ships but, after being attacked by the carrier's aircraft, was forced down into the open sea where its surviving crewmembers were picked up by the Japanese and became POWs.

A more successful engagement occurred soon after, when on a night raid to Rabaul, Squadron Leader Hemsworth's Catalina was attacked and damaged by three fighters. Sergeant D Dick, an air gunner on his first mission, shot down one of the enemy aircraft.

By early 1942 air raids against Port Moresby had become both frequent and heavy. Despite the destruction of facilities and flying boats, operations which now included convoy escorts and supply dropping, continued until May when the unit moved to Bowen. From here patrolling Catalinas made important sightings of Japanese shipping in the lead up to the battle of the Coral Sea. A further move to Cairns was effected in October.

Despite these moves, 20 Squadron continued to operate extensively over New Guinea conducting night bomber raids. Many of these missions were flown through intense tropical storms which would have daunted all but the staunchest crews. Dangerous flying conditions created by the weather were probably responsible for the loss of several Catalinas.

1943 saw the addition of yet another task, that of mine laying. In this role the Catalinas operated throughout the Pacific, but particularly against targets in the Dutch East Indies. From October 1944, 20 Squadron was based at Darwin and aircraft detailed for mining duties often staged through Moratai to reduce the distance from their targets. Later in 1945, detachments operated from the recaptured Philippines and made the RAAF's most northern penetrations of the Pacific war when they mined harbours along the Chinese coast.

As 1945 progressed the Catalinas also flew some night bombing sorties in support of the Australian invasion of Labuan. On July 30 1945, 20 Squadron conducted the RAAF's last mine laying operation when three aircraft mined Banka Strait.

Following Japan's surrender, 20 Squadron for a time evacuated Australian POWs from Singapore prior to moving to Rathmines in November where it disbanded on March 27 1946.

A Catalina performs a rocket assisted takeoff.

Formed: August 1 1941
Squadron Code: RB
Aircraft: Empire, Catalina
Locations: Port Moresby: Aug 1941 – May 1942
Bowen: May – October 1942
Cairns: October 1942 – October 1944
Darwin: October 1944 – November 1945
Rathmines: November 1945 – March 1946
Disbanded: March 27 1946

21 SQUADRON

As an element of the Citizen Air Force, 21 Squadron formed at Laverton on April 20 1936.

Commanded by Squadron Leader J H Summers, the unit operated Hawker Demons, Westland Wapitis and de Havilland Gipsy Moths and comprised both reservists and regular personnel.

With the types of aircraft held, training was varied and consisted of bomber, fighter and army co-operation work. Searches for missing aircraft and small surface craft were conducted and the unit also participated in a number of airshows. In December 1936 the first Anson Mk.Is were received while in January 1937 four Bulldog IIA fighters were allocated.

On September 3 1939, 21 Squadron mobilised for war. Operating Demons, Ansons, Avro Cadets and a North American NA-16 (the forerunner of the CAC Wirraway), training of personnel was intensified. The first operational mission, an uneventful convoy escort patrol, was undertaken on December 15.

Re-equipped with Wirraways and standardising in an army co-operation role, 21 Squadron moved to Seletar on Singapore in August 1940. A further move to Sembawang was effected in February 1941 where the unit re-equipped with Buffalo Mk.Is and became a fighter squadron.

With its new aircraft, 21 Squadron moved to Sungai Pattani where it was destined to be the only fighter squadron based on the Malayan mainland when Japan entered the war on December 7 1941. The next day, Sungai Pattani was repeatedly attacked by Japanese aircraft and several Buffaloes were destroyed or damaged. Vital airfield facilities were also wrecked. After unsuccessful attempts to intercept enemy bomber and fighter formations, 21 Squadron vacated Sungai Pattani and moved to Butterworth and then Ipoh as continuing large scale Japanese attacks destroyed facilities and aircraft at these bases. It was during an attack against Butterworth that Flying Officer H V Montefiore obtained 21 Squadron's first victory when he shot down a Zero fighter.

Despite receiving replacement Buffaloes, intensive Japanese attacks continued to take a heavy toll on the Australian squadron. Reserves of Buffaloes were soon exhausted and 21 Squadron merged with 453 Squadron (which also operated Buffaloes) at Kuala Lumpur on December

25. The combined unit was provisionally known as 21/453 Squadron.

By the end of December, the unit had been forced back to Singapore by rapidly advancing Japanese ground forces. From Sembawang the few remaining Buffaloes escorted convoys bringing reinforcement to Singapore, carried out tactical reconnaissances, tried to intercept incoming attacks and escorted bombers attacking enemy positions.

In a campaign notable for disasters, 21/453 Squadron flew a successful mission on January 16 1942 when six Buffaloes escorted Vilderbeests on a bombing mission over the Malayan mainland. Three enemy fighters attacked the bombers but before they could cause any damage they were all probably destroyed by the Buffaloes. The 'kills' could not be confirmed as it was necessary for the fighters to remain with the bombers.

Not so successful were attempts to intercept Japanese bomber formations which, by now, were regularly attacking targets (including Sembawang) on Singapore Island. Warning time was simply insufficient to enable the Buffaloes to reach the enemy's height before they attacked.

On January 26 all available Buffaloes escorted bombers sent to strike at an invasion force landing at Endau. Two Japanese fighters were destroyed and another probably destroyed in these missions. On the same day, 21 Squadron was separated from 453 Squadron, handed over its remaining aircraft and equipment to that unit and was transferred to Sumatra and later Java. There its personnel were employed on aircraft servicing tasks and general airfield duties prior to being evacuated to Australia.

During the last year of World War II, 21 Squadron operated Liberator heavy bombers against a variety of Japanese targets. (RAAF Museum)

A 21 Squadron Wirraway.

After its arrival in Australia, 21 Squadron was reduced to a nucleus until re-established at Gawler with Vengeance dive bombers in September 1943. Two months later, the unit moved to Lowood and in early 1944 to Nadzab where combat operations commenced on February 22. On this occasion three 21 Squadron aircraft joined others from 23 and 24 Squadrons in a joint attack against barge hideouts on the Wagor River, Madang. Next day, another combined attack was launched against supply dumps near Erima jetty and troop concentrations in the vicinity of Singor village. Over succeeding days airfields, barge traffic and Japanese forward positions came under very accurate attack by the Vengeances. Unfortunately, after only a few weeks, the Vengeances were withdrawn from New Guinea on the orders of the Commanding General, Allied Air Forces. 21 Squadron relocated to Camden in March where flying was limited to army and naval co-operation work. By June the Vengeances had been withdrawn and the unit moved to Leyburn where it converted to Liberator heavy bombers.

After its conversion training had been completed, 21 Squadron moved to Fenton and recommenced operations on January 11 1945. These were mainly confined to armed reconnaissance sorties, which sometimes resulted in bombing and strafing attacks against barges and small ships. Occasionally the Liberators were attacked by Japanese fighters, however these combats were usually inconclusive.

On April 6, 21 Squadron Liberators participated in a major attack against a Japanese convoy comprising the cruiser *Isuzu* and four escorting vessels. Despite fighter opposition, which shot down two 24 Squadron Liberators, and anti-aircraft fire the bombers pressed home their attacks but failed to secure any hits on the ships, some of which were later sunk by an American submarine.

By this time, the unit had a permanent detachment operating from Moratai to support the invasion of Tarakan. On August 12, 21 Squadron flew its last wartime operation when three Liberators bombed Halmahera. Just days later Japan surrendered and the Liberators were then used to ferry personnel from the islands to Australia.

In December 1945, 21 Squadron moved to Tocumwal and in April of the following year to Amberley where it re-equipped with Lincolns in 1947. For a time the unit still maintained a specially converted Liberator for use as a VIP transport aircraft. On February 23 1948, 21 Squadron was renumbered 2 Squadron.

21 (City of Melbourne) Squadron reformed at Laverton on April 1 1948. Initial aircraft equipment comprised Mustang, Wirraway and Tiger Moth aircraft. As before the war, general training and exercises involving both regular and reserve force personnel were undertaken and, in January 1951, the first cadet pilot training course commenced.

During July 1951 Vampire jets were received as replacements for the Mustangs, although the last Mustang was not retired until late 1955. Later, a Sikorsky S-51 helicopter was added to the unit's fleet and in 1959 Australian built Winjeels replaced the Wirraways.

21 Squadron ceased operations as a flying unit on June 2 1960 and commenced base support duties. Still at Laverton, 21 Squadron continues to provide a RAAF Reserve support component to the Permanent Air Force in the Melbourne area.

Formed: *April 20 1936*
Squadron Code: *R, GA, MJ*
Aircraft: *Demon, Wapiti, Gipsy Moth, Anson, Cadet, NA-16, Wirraway, Buffalo, Vengeance, Liberator, Lincoln, Mustang, Tiger Moth, Vampire, Winjeel*
Locations: *Laverton: April 1936 – August 1940*
Selatar: August 1940 – February 1941
Sembawang: February – November 1941
Sungai Pattani: November – December 1941
Butterworth: December 1941
Ipoh: December 1941
Sembawang: December 1941 – February 1942
Palembang: February 1942
Batavia: February – March 1942
Disbanded:
Gawler: September – November 1943
Lowood: November 1943 – January 1944
Nadzab: January – March 1944
Camden: March – July 1944
Leyburn: July 1944 – January 1945
Fenton: January – July 1945
Balikpapan: July – December 1945
Tocumwal: December 1945 – April 1946
Amberley: April 1946 – February 1948
Disbanded:
Laverton: April 1948 – present

---

22 SQUADRON

Equipped with Hawker Demons and de Havilland Gipsy Moths, 22 Squadron formed at Richmond on April 20 1936. Under the command of Squadron Leader D E L Wilson, the new squadron, like its sister Citizen Air Force (or as they are now known Reserve squadrons), staff comprised two thirds of reserve and one third regular personnel.

Up until the outbreak of World War II, 22 Squadron trained in a general purpose role and as a subsidiary task trained its own cadet pilots to wings standard. Anson Mk.Is joined the unit from early 1937.

Upon the declaration of war with Germany 22 Squadron was mobilised and flew its first operational mission – a search off Sydney for a reported submarine – on September 10. As, however, there was almost no enemy activity in the South West Pacific area until 1942, aside from some anti-submarine and convoy escort patrols,

22 Squadron's most famous type – the Boston. It was in one of these aircraft that Flight Lieutenant W E Newton was awarded the Victoria Cross.

little out of the routine occurred. During December 1940 Australian built CAC Wirraways began to supplement and then replace the Demons and, by December 1941, the unit was fully equipped with these aircraft.

From April 1942, 22 Squadron began to convert to American built Douglas A-20 Boston attack bombers. These particular aircraft had been ordered by the Dutch, but upon the fall of the East Indies they were diverted to Australia where they found their way into 22 Squadron. They came without spare parts, tools and essential support equipment and to make matters worse, technical instructions and other documentation were printed in Dutch. Under these circumstances the conversion proved to be a most difficult process.

Despite these difficulties, 22 Squadron was exceedingly fortunate to receive these superlative aircraft which were among the very best in Australia at that time. While still converting to its new aircraft, 22 Squadron was called upon to conduct anti-submarine patrols off Australia's south eastern coast. Such patrols were intensified after the Japanese midget submarine attack on Sydney Harbour.

On June 6 Flying Officer V W Morgan's crew attacked a submarine, while four days later Flying Officer J C Mill's crew also made an attack. In both cases large quantities of oil came to the surface and it was thought both vessels had been sunk. It was found after the war, however, that both submarines had escaped destruction.

In October 1942, 22 Squadron moved to Wards strip, Port Moresby. Flying in New Guinea, however, began disastrously when Flight Lieutenant Morgan's crew were killed after their Boston exploded during bombing practice. On November

15, combat operations commenced with the Bostons flying in close support of Australian troops near Port Moresby and attacking supply lines and reinforcements behind enemy lines.

Unfortunately, early operations were marred by further mysterious losses. On November 26, while five Bostons were attacking Buna airfield, Squadron Leader K McDonald's aircraft exploded and just three days later, Flight Lieutenant J Bullmore's aircraft was lost in similar circumstances during an attack on Gona.

An investigation later revealed that all three aircraft – including Flight Lieutenant Morgan's – had been destroyed by the small 9.8kg fragmentation bombs they had been carrying which had detonated prematurely in the aircraft's slipstream. Discontinuing the use of these bombs, and the subsequent end to this type of loss, confirmed the theory that this was indeed the cause of the problem.

By day, and occasionally by night, attack operations continued at an intensive pace and from December many of these missions were flown in conjunction with the Beaufighters of 30 Squadron. The Australian Bostons ranged over all of New Guinea flying their low level attack missions with bombs and machine guns as their offensive weapons.

In mid December the Bostons achieved an important success when they badly damaged a Japanese destroyer which formed part of a convoy heading for Rabaul. The ship, crippled by two very near misses, was later sunk by American B-17 Flying Fortresses and its destruction was jointly credited to both air forces.

The first quarter of 1943 saw a continuance of 22 Squadron's low level strikes – particularly in support of the Allied drive to Buna. Attacks against well defended Lae airfield were also a feature of the unit's work. These operations, which were pressed on regardless of opposition, were quite successful and resulted in the destruction of aircraft and facilities. On some occasions the Bostons were intercepted by fighters, but with their excellent manoeuvrability and good turn of speed, invariably escaped from these encounters.

Anti-aircraft fire was the main hazard and during one raid no less than three Bostons were damaged. One had an anti-aircraft shell explode in its cockpit which wounded the pilot, Flight Lieutenant P C Mullens, in the face. Despite his injuries, Mullens continued his attack run and later made a safe landing at Port Moresby.

On March 2, 22 Squadron played an important part in the Battle of the Bismarck Sea – one of the pivotal actions of the South West Pacific campaign – when six Bostons attacked Lae airfield to temporally neutralise this vital airfield during the coming air-sea engagement. A second strike was conducted in the afternoon.

As dawn broke the following morning the Bostons again swept in low over Lae. The runways had been repaired overnight and Japanese fighters were preparing to takeoff to escort the approaching Japanese ships into Lae harbour when they were caught on the ground by the Australian aircraft. Several were destroyed by strafing and bombing while once again the runways were cratered and made unserviceable. A follow up attack that afternoon ensured that the Japanese pilots could not support their naval comrades who were already under attack.

After participating in yet another strike against Lae airfield on the morning of March 4, 22 Squadron itself was committed to making a low level skip bombing attack on the Japanese convoy. It is now history that all eight Japanese transports and four of the eight escorting destroyers were sunk in this battle but it is not generally realised that 22 Squadron played an important part in the sinkings. Despite anti-aircraft fire and the presence of Zero fighters which attacked repeatedly, the Bostons completed their attack runs from wave top height. Two direct hits and a further ten near misses were scored and all the Bostons returned safely from the engagement.

Just two weeks later on March 16 six Bostons attacked buildings and newly constructed oil storage tanks at Salamaua. Flying through exceedingly heavy anti-aircraft fire, the Bostons pressed onto their allotted targets, Flight Lieutenant W E Newtons crew destroying two oil tanks before their aircraft was rocked by explosions. Hit by no less than four anti-aircraft shells, the Boston suffered damage to both engines, control surfaces, fuel tanks, electrical, hydraulic and fuel systems. Despite its battered condition, Newton bought his aircraft back to base and

managed to land without injury to his crew. For his heroic actions on this day Newton was later awarded the Victoria Cross – the highest gallantry award available to a member of the British Commonwealth and the only such award made to a member of the RAAF in the Pacific theatre.

Six Bostons again attacked Salamaua on March 18. Newton was again flying and again hit by anti-aircraft fire. On this occasion his aircraft burst into flames just as he destroyed his target with a direct hit. There was no chance of keeping the aircraft airborne and Newton put the Boston down onto the sea near Salamaua where he and one of his crewmen – Flight Sergeant J Lyon – managed to swim ashore where they were captured. The Boston's other crewman, Sergeant B G Eastwood, did not leave the aircraft and may have been killed by anti-aircraft fire. After being interrogated by their captors, both airmen were executed – Flight Sergeant Lyon was bayoneted to death at Lae while Flight Lieutenant Newton was beheaded at Salamaua.

22 Squadron moved to Goodenough Island in July 1943. From here attacks of the type already described continued – at times in the company of Beaufighters, Beauforts and Kittyhawks. During November the unit again moved – this time to Kiriwina. Attack operations and barge sweeps designed to further isolate Cape Gloucester were conducted and another regular task was the bombing of the almost derelict Gasmata airfield.

Some losses to anti-aircraft fire were experienced while conducting these tasks and, not for the first time, the unit had difficulty in maintaining a sufficient number of aircraft available for operations. Targets in New Britain were later added to 22 Squadron's responsibility and a number of photographic reconnaissance sorties were also flown.

In mid November 1944, 22 Squadron commenced operations from Moratai. The Bostons ranged over their island targets up to and including the Philippines, however, just weeks later, the unit was dealt a severe blow when, on the night of November 22, eight Japanese aircraft bombed the Australian's base. Although 22 Squadron escaped any casualties amongst its personnel, several bombs landed amongst its closely parked aircraft and four Bostons were totally destroyed and another seven so badly damaged that they had to be written off. With no replacements available operations were severely curtailed until the unit re-equipped with Australian built Bristol Beaufighters at Noemfoor in January 1945.

Beaufighter operations commenced on February 13 when nine aircraft participated in a massed attack against Tandano in the Celebes. With their heavy armament of 20mm cannon, machine guns, rockets and bombs, the Beaufighters participated in many attack missions and barge sweeps during the year. While Japanese air opposition had long been non existent, anti-aircraft fire was still a deadly hazard and several Beaufighter crews were lost in these low level attacks.

From April attacks in support of the impending invasion of Tarakan were conducted and the following month after the invasion 22 Squadron began to move to Tarakan. The airfield allotted, however, was not usable and the unit remained split with most of its ground personnel at Tarakan and the aircraft and aircrew at Moratai.

Attack operations continued up to the Japanese surrender after which the Beaufighters dropped surrender leaflets over prison camps and Japanese occupied territory. 22 Squadron returned to Australia in December 1945 and disbanded at Deniliquin on August 15 1946.

22 (City of Sydney) Squadron reformed at Bankstown on April 19 1948. As had been the case prior to World War II, the reformed unit contained a nucleus of Permanent Air Force personnel while the remainder were Citizen Air Force volunteers.

In November, the unit moved to Schofields where it received its first North American Mustangs and de Havilland Tiger Moths. With these aircraft it trained its regular and Citizen Air Force pilots to a high level of proficiency and participated in air defence, naval co-operation and army support exercises prior to relocating to Richmond in March 1953. For a time 22 Squadron operated a Sikorsky S-51 Dragonfly rescue helicopter but the unit's most important addition

occurred in September 1952 when the first Australian built FB.30 and T.33 de Havilland Vampire jet fighters were taken on charge. The conversion was a slow process however and both Vampires and Mustangs were operated side by side for two years with the Mustangs gradually being relegated to a secondary role.

In April 1956 Gloster Meteor F.8s began to replace the Vampires although both types remained in service until June 1960 when Citizen Air Force flying operations ceased. Today, a non flying 22 Squadron continues to support Permanent Air Force activities from its base at Richmond.

Formed: April 20 1936
Squadron Code: S, DU
Aircraft: Demon, Gipsy Moth, Anson, Wirraway, Boston, Beaufighter, Mustang, Tiger Moth, Dragonfly, Meteor, Vampire, Winjeel
Locations: Richmond: April 1936 – Oct 1942
Port Moresby: October 1942 – July 1943
Goodenough Island: July – November 1943
Kiriwina: November 1943 – August 1944
Noemfoor: August – November 1944
Moratai: November – December 1944
Noemfoor: December 1944 – February 1945
Moratai: February – May 1945
Tarakan: June – July 1945
Moratai: July – December 1945
Deniliquin: December 1945 – August 1946
Disbanded:
Bankstown: April – November 1948
Schofields: November 1948 – March 1953
Richmond: March 1953 – present

---

23 SQUADRON

23 Squadron first formed at Laverton on May 3 1937. Equipped with Hawker Demon Mk.Is and Avro Cadet Mk.II biplanes, the unit, under the command of Wing Commander R J Brownell – an ace fighter pilot of World War I – moved to Perth in March 1938. After receiving Anson Mk.Is, the unit was renumbered 25 Squadron on January 1 1939.

Again equipped with Ansons, 23 Squadron reformed at Richmond the following month. As a General Purpose Citizen Air Force squadron, the unit's 300 personnel were drawn from both the Regular and Citizen Air Force.

Early operations mainly consisted of naval co-

operation flying off the east coast of Australia. In August the unit moved to Archerfield, near Brisbane, where it re-equipped with Wirraways and commenced seaward reconnaissance and anti-submarine patrols after the commencement of hostilities with Germany.

In August 1940, 23 Squadron received the first of several Lockheed Hudsons. With these more capable aircraft and its CAC Wirraways, the unit spent the remainder of 1940 and all of 1941 engaged in uneventful seaward patrols and convoy escort missions. On March 24 1942, 23 Squadron's first attack was made when Sergeant G Herring, in a Wirraway, dropped two bombs on

23 Squadron's Citizen Air Force personnel flew and maintained Australian built Mustangs during the late 1940s and 1950s.

a submarine off Brisbane. A brown substance came to the surface but the attack, which is believed to have been the first off Australia's east coast, could not be claimed as successful.

23 Squadron moved to Amberley in May but the following month it relocated to Lowood. Within weeks of this move, a number of Bell P-39D Airacobra fighters were taken on strength. The Hudsons had by now left and the unit's operational strength stood at six Airacobras and 18 Wirraways.

Operations continued without major incident until June 1943 when 23 Squadron's role was changed to that of a dive bomber unit and the first Vengeance aircraft began to arrive. When fully rearmed, 23 Squadron moved to Nadzab in February 1944. After some familiarisation flying, operations commenced on February 11 when six Vengeances, along with another six from 24 Squadron, struck targets around Saidor in support of American ground forces. The resulting bombing was extremely accurate and drew the excited praise of the Americans.

Strikes continued throughout the month, with particular emphasis on targets at Alexishafen. As with its first operation, these attacks resulted in very accurate bomb deliveries. Despite its obviously successful attacks, 23 Squadron along with other RAAF Vengeance squadrons, was withdrawn to the Australian mainland on the orders of General Kenney, the commander of Allied Air Forces. The Vengeances, he contended, were obsolete and better aircraft could be employed more productively in forward areas where airfield space was at a premium. After flying its last mission on March 8 1944, 23 Squadron found itself languishing at Higgins Field on Cape York Peninsula.

23 Squadron later moved to Menangle, where it was reduced to cadre pending re-equipment with Consolidated B-24 Liberator heavy bombers. By November the unit had moved again, this time to Leyburn. After taking delivery of its Liberators, 23 Squadron deployed to Long in the Northern Territory during April 1945 and commenced operations on April 7 when two Liberators were despatched on anti-shipping patrols. Until the end of the war operations consisted of reconnaissances, strikes on shipping and occasional bombing attacks, particularly against targets on Timor and the East Indies.

A very successful strike was conducted when six Liberators struck Maomere, a port in the Flores Islands on May 10. Around 20 large fires and explosions resulted and a number of boats were destroyed. Another successful raid occurred on June 2 when the Liberators destroyed four camouflaged Japanese aircraft at Cape Chater airfield on Timor.

23 Squadron began a temporary move to Darwin in June 1945, and became involved in the preparations for the invasion of Borneo. From June 10, using Darwin and Moratai as bases, the unit struck at targets on Borneo. The ground staff embarked in a Liberty ship which offloaded them at Moratai. From here they sailed for Balikpapan, arriving on July 16.

After Japan's surrender supply dropping flights to Allied POWs still in Japanese hands were made. The task of evacuating released POWs and other personnel to Australia then commenced. To facilitate this work the ground staff moved from Balikpapan to Moratai where the Liberators were then based.

23 Squadron moved to Tocumwal in November 1945, however the Liberators were still employed evacuating personnel from New Guinea and Moratai. By 1946 this work had largely been completed and 23 Squadron moved to Amberley

where it was renumbered 6 Squadron in February 1948.

On April 1 1948, 23 (City of Brisbane) Squadron reformed at Archerfield. Equipped with Tiger Moths, Wirraways and Mustangs, the unit, as a Citizen Air Force squadron, commenced its own Cadet pilot training program later in the year and participated in various exercises.

23 Squadron moved to Amberley in September 1955 and re-equipped with de Havilland Vampire jet fighters. From October 1956 these were supplemented, and later replaced by Gloster Meteor F.8s while Winjeels replaced the Wirraways in 1959. Flying activities ceased in June 1960 and the unit assumed a ground training role.

Formed: *May 3 1937*
Squadron Code: *T, NV*
Aircraft: *Demon, Cadet, Anson, Wirraway, Hudson, Airacobra, Vengeance, Liberator, Mustang, Tiger Moth, Dragonfly, Vampire, Meteor*

Locations: *Laverton: May 1937 – March 1938*
Pearce: March 1938 – January 1939
Disbanded:
Richmond: February – August 1939
Archerfield: August 1939 – May 1942
Amberley: May – June 1942
Lowood: June 1942 – February 1944
Nadzab: February – March 1944
Higgins Field: March 1944 – July 1944
Menangle: July – October 1944
Leyburn: October 1944 – March 1945
Long: April – June 1945
Darwin: June – July 1945
Balikpapan: July – November 1945
Tocumwal: November 1945 – April 1946
Amberley: April 1946 – February 1948
Disbanded:
Archerfield: April 1948 – September 1955
Amberley: September 1955 – present

---

24 SQUADRON

Commanded by Squadron Leader C A Campbell and equipped with CAC Wirraway and de Havilland Moth Minor aircraft, 24 Squadron formed at Amberley on June 17 1940. The unit moved to Townsville in October where a number of Lockheed Hudson bombers were taken on charge. General purpose training and maritime patrols were conducted with little incident until early December 1941 when 24 Squadron moved to Rabaul.

Upon Japan entering the war, patrols were intensified and, as Japanese forces came closer attacks were mounted on their bases. Rabaul itself, however, soon came under attack from Japanese aircraft although operations continued until February 20 1942 when over 100 Japanese aircraft struck at the island. Eight Wirraways took off to intercept the raiders but were immediately attacked by a considerable force of Zero fighters. Against such odds no effective defence could be offered and three Wirraways were shot down, two crash landed and another was damaged. The Japanese suffered no losses and proceeded to bomb the airfield and other facilities against no opposition.

With only two Wirraways and one Hudson remaining, the next morning 24 Squadron was ordered to attack a Japanese invasion force comprising two aircraft carriers, cruisers, destroyers and transports. Clearly such an operation was suicidal and had no chance of success and, on receipt of this order, Wing Commander J Lerew signalled RAAF Headquarters his now famous Roman gladiators' salutation "Moriturb Vos Salutamus" ("we who are about to die salute you"). A nonplussed RAAF Headquarters heeded the message and, instead of insisting that the attack proceed, withdrew the unit to Townsville.

Replacement Wirraways and crews were received as was an impressed Junkers and two Ford Trimotor transports. These aircraft though

24 Squadron Vengeance dive bombers return from a strike against Japanese positions at Shaggy Ridge in January 1944. 24 Squadron operated the Vengeance in active operations longer than any other RAAF unit.

24 Squadron Liberators running in to bomb Laha airfield in the Dutch East Indies. (RAAF Museum)

convoy. The convoy, which had a fighter escort, escaped damage and several Australian aircraft, including two 24 Squadron Liberators, were shot down.

Strikes, particularly against Balikpapan, continued until Japan's surrender in August, whereupon the Liberators were used to ferry POWs and other personnel from Moratai to Australia. By the end of 1945, 24 Squadron had relocated to Tocumwal, where it disbanded on May 15 1946.

24 (City of Adelaide) Squadron reformed at Mallala on April 30 1951. Equipped with Mustangs, Wirraways and later Tiger Moths, the unit was allocated a fighter role. General training and exercises, often centring on weekends, were conducted and cadet pilots were trained to wings standard. In January 1959, two Winjeel trainers were received but, in June 1960, the unit ceased operations as a flying squadron and moved to Edinburgh where it commenced, and continues to operate in, a base support role. 24 Squadron was the only Citizen Air Force unit to fly Mustangs for the entire postwar period – the other squadrons all having converted to jets.

were only operated for a short time and were handed over to other units.

In July 24 Squadron moved to Bankstown where it was allocated five Bell Airacobra fighters in addition to its Wirraways. General training, anti-submarine patrols and interceptions of unidentified aircraft were conducted until mid 1943. This strange assortment of aircraft increased in May when the first Vultee Vengeance dive bombers were received and in June three Buffalo fighters were allocated.

By August, the unit had standardised on Vengeances and proceeded to New Guinea where dive bomber operations in support of the Australian Army were conducted from Nadzab. Accurate attacks were made against enemy occupied towns and on Japanese positions at Shaggy Ridge. The Vengeances also supported the Cape Gloucester landings before being withdrawn from New Guinea in March 1944. Of all the RAAF's Vengeance units, 24 Squadron operated the type far longer in action than the others and obtained very good results with this much maligned aircraft.

Moving to Lowood, 24 Squadron re-equipped with Consolidated Liberator heavy bombers and then proceeded to Manbulloo in the Northern Territory where operations commenced in September after a further move to Fenton. From here, anti-shipping strikes, armed reconnaissance missions and bomber attacks against enemy occupied territory were conducted.

On March 6 1945, 24 Squadron participated in an unsuccessful formation attack on a Japanese

Formed: June 17 1940
Squadron Code: U, GR
Aircraft: Wirraway, Hudson, Airacobra, Buffalo, Junkers W34F trimotor, Battle, Ford trimotor, Vengeance, Liberator, Mustang, Tiger Moth, Winjeel
Locations: Amberley: June – October 1940
Townsville: October 1940 – December 1941
Rabaul: December 1941 – January 1942
Townsville: January – July 1942
Bankstown: July 1942 – August 1943
Nadzab: August 1943 – March 1944
Lowood: March – June 1944
Manbulloo: June – September 1944
Fenton: September 1944 – July 1945
Moratai: July – December 1945
Tocumwal: December – May 1946
Disbanded:
Mallala: April 1951 – March 1960
North Adelaide: March – May 1960
Edinburgh: May 1960 – present

25 SQUADRON

On January 1 1939, 23 Squadron, based at Pearce, was renumbered 25 Squadron. Equipped with Avro Ansons, Hawker Demons and Avro Cadets trainers, the Citizen Air Force units role was that of a general purpose squadron.

Until September 1939, 25 Squadron, under the able command of Wing Commander R S Brownell, was mostly engaged in army support tasks, naval co-operation work, meteorological flying and cadet pilot training. Several de Havilland Moth Minors were received during this period and were used for pilot training and some communications flying. After war was declared, all reserve personnel were mobilised and Australian built CAC Wirraways replaced the Demons. The Wirraways, frequently detached to other Western Australian bases, conducted anti-submarine patrols and convoy escort missions over a wide area.

Active service eluded 25 Squadron throughout 1941 although the unit's aircraft were used to search for Australian and German survivors off Carnarvon after a naval engagement between the cruiser HMAS *Sydney* and the German surface raider *Kormoran*. Both ships were sunk in the action and, while none of *Sydney's* complement were recovered, German survivors were located and subsequently rescued.

With Australia's dramatically changed circumstances resulting from Japan entering the war, March 1942 found 25 Squadron's Wirraways supplemented with three Fairey Swordfish. While the sight of these ungainly biplanes would hardly have reassured Perth's population as to their security, before being withdrawn some months later, they proved very useful in the anti-submarine patrol role. 1942 also saw the allocation of a small number of Brewster Buffalo Mk.Is and with these obsolete aircraft 25 Squadron was charged with the air defence of Perth – a tall order for a handful of Buffaloes and Wirraways. It is undoubtedly fortunate that Perth was never attacked by the Japanese.

After receiving a small allocation of Vultee Vengeances late in 1942, by February 1943, 25 Squadron was still equipped with nothing more impressive than Wirraways. Plans were in hand to fully re-equip with Vengeances, however, and by August 1943 the unit was fully armed with these dive bombers. Due to the almost nonexistent Japanese activity around Western Australia and its adjacent waters, these new aircraft were used for nothing more than anti-submarine patrols and army co-operation exercises.

In January 1945 after relocating to Cunderdin, 25 Squadron began to rearm with Liberator heavy bombers. The arrival of these new long range aircraft at last offered the unit's personnel the opportunity to play an active part in the war. Even before finishing their conversion training some of the Liberator crews undertook long range searches for the survivors from a merchant vessel sunk by the German submarine U-862 1300 kilometres off Fremantle. Survivors located during these flights were later rescued by ships directed into the area.

25 Squadron's first bombing mission was flown on March 13 1945 when six Liberators staged through Truscott to bomb Japanese barges and base facilities at Mapin on Sumbawa Island. Up until the end of the war, 25 Squadron flew long range missions striking at targets in the Dutch East Indies. From September 1945 to January 1946, 25 Squadron aircraft evacuated recently released POWs from the islands to Australia. This proved to be 25 Squadron's last major wartime task and the unit disbanded at Cunderdin on July 9 1946.

25 (City Of Perth) Squadron reformed at Pearce on April 1 1948. Equipped with Mustangs, Wirraways and Tiger Moths, the units role was to train cadet pilots and, with its regular and reserve personnel, to maintain a RAAF fighter presence in Western Australia. In September 1951, the unit received its first Vampire jet fighter but both Vampires and Mustangs operated side by side until early 1956 when the jets completely replaced the reliable piston engined aircraft.

For a time 25 Squadron operated the obsolete Brewster Buffalo and was tasked to provide the air defence of Perth. Fortunately they did not see combat.

(right) As a Citizen Air Force unit, 25 Squadron operated Vampires postwar. (via RAAF Museum)

25 Squadron continued to operate as a jet fighter squadron until June 1960 when it, along with the other Citizen Air Force Squadrons, ceased flying and commenced a ground training role. For the next 30 years 25 Squadron provided ground support for RAAF flying squadrons and later base support functions at Pearce until November 1 1989 when it again reverted to a flying role with Macchi MB-326 jets.

Today, 25 Squadron is unique in the RAAF. The only reserve unit with operational aircraft, its role is to provide lead in training for recently graduated pilots allocated to either Hornet of F-111 aircraft. Coupled with this task, the Macchis also carry out fleet support duties in Western Australian waters. Occasional Army close support missions are also flown. While the Macchis have only limited utility in the latter role, a replacement is due to enter service by the end of the decade. Any new aircraft will undoubtedly have a more useful ground attack capability – a capability sorely needed particularly in support of the Army. 25 Squadron's future seems assured and the unit is expected to continue in its allocated role for some years to come.

Formed: *January 1 1939*
Squadron Code: *V, SJ*
Aircraft: *Anson, Demon, Cadet, Wirraway, Moth Minor, Swordfish, Buffalo, Vengeance, Liberator, Mustang, Tiger Moth, Vampire, Winjeel, Macchi*
Locations: *Pearce: Jan 1939 – Jan 1945*
Cunderdin: January 1945 – July 1946
Disbanded:
Pearce: April 1948 – present

30 SQUADRON

Equipped with Bristol Beaufighters and allotted a long range attack role, 30 Squadron formed at Richmond under the command of Squadron Leader C Read on March 9 1942.

Due to the extent of recent Japanese successes, training was necessarily hurried and in mid August 30 Squadron moved to Bohle River. On September 6 three Beaufighters deployed to Milne Bay where, next day, they attacked Japanese shipping. This strike, flown by two of the three aircraft, was the first RAAF Beaufighter combat operation of the Pacific war.

During September the rest of 30 Squadron moved to Port Moresby and low level strikes against Japanese troop concentrations and seaborne barge traffic commenced. In both roles, the Beaufighters, with their heavy cannon and

machine gun armament, proved particularly effective. Japanese airfields, especially Lae, also came in for damaging attacks. Conditions at Port Moresby in these dangerous days were quite primitive thus making aircraft maintenance tasks far more difficult than on mainland Australia. The unhealthy tropical climate combined with poor domestic conditions ensured that air and ground crews began to suffer the effects of tropical illness soon after arriving in New Guinea.

It was not long before the Australians suffered their first losses when, on September 23 Flying Officer G W Sayer's Beaufighter was hit by anti-aircraft fire over Buna and crashed. Operating at low level – sometimes extreme low level – left the Beaufighter crews almost no time to escape from their crippled aircraft. Other losses in similar circumstances followed and remained a hazard of Beaufighter operations up until the end of the war.

On March 3 1943, 30 Squadron participated in the Battle of the Bismarck Sea. In this engagement the Beaufighters attacked a 16 ship Japanese convoy from mast height to suppress anti-aircraft fire. The Japanese, under the mistaken impression that they were under torpedo attack, made a disastrous tactical error and turned their ships towards the Beaufighters, leaving them exposed to attack by American low level bombers. Eight troop laden transports and four destroyers were sunk in this battle for the loss of five aircraft, including one Beaufighter.

Damien Parer – the famous war photographer – filmed part of this gripping attack from Flight Lieutenant R F Uren's aircraft. This footage subsequently received wide circulation in Australia and overseas.

The day after the battle 30 Squadron again visited Lae where they caught the base defenders unprepared. Aside from the destruction of facilities, the Beaufighters destroyed six Zeros on the ground and killed a number of personnel.

30 Squadron moved to Goodenough Island in July where the Beaufighters' main tasks were attacks against barges and airfields. The airfields in particular were well defended by anti-aircraft batteries and fighters, consequently, while the Beaufighters inflicted destructive losses on the Japanese, 30 Squadron, at times, also suffered heavily. In November the unit relocated, this time to Kiriwina.

June 1944 saw 30 Squadron again on the move, this time to Tadji where Beaufighter attacks proved so successful that Japanese barge traffic was quickly halted in daylight. Air-to-ground rockets were added to the Beaufighter's armament in October and the following month 30 Squadron moved to Moratai. From here, attacks were conducted against targets in the Celebes, Ambon, Ceram and the Halmaheras.

30 Squadron itself came in for some unwelcome attention from the enemy on one occasion whilst at Moratai. In a small scale night attack, Japanese bombers destroyed two Beaufighters and damaged a further eight. The damaged aircraft were quickly repaired or replaced and operations resumed at their previous rate.

May and June 1945 saw 30 Squadron in the process of moving to Sanga Sanga to support the assault on Tarakan but this had not been completed by the end of hostilities. 30 Squadron disbanded at Deniliquin on August 15 1946.

30 Squadron Beaufighters at their tropical lair. This unit operated British and later Australian built Beaufighters with deadly effect throughout World War II. (RAAF Museum)

On March 3 1948 the Target Towing and Special Duties Squadron was redesignated 30 Squadron. Equipped primarily with Beaufighters but also operating Beauforts, Dakotas, Wirraways, Ansons and Mustangs, the unit operated in the target towing role from Richmond. A series of moves in this period took 30 Squadron to Schofields, Fairbairn and then back to Richmond. Civil aid tasks were also carried out until it disbanded on March 21 1956.

Equipped with Bloodhound Mk.I surface-to-air missiles, 30 Squadron reformed for the last time at Williamtown on January 11 1961 under the command of Squadron Leader E W Tonkin. The role of the newly formed missile squadron was to provide high level air defence of Australian military bases and industrial centres. From 1965 a permanent detachment was based at Darwin. 30 Squadron, which had the distinction of being the RAAF's only surface-to-air missile unit, disbanded on December 2 1968.

Formed: *March 9 1942*
Squadron Code: *LY*
Aircraft: *Beaufighter, Beaufort, Dakota, Wirraway, Anson, Mustang, Bloodhound (SAM)*
Locations: *Richmond: March – August 1942*
Bohle River: August – September 1942
Port Moresby: September 1942 – July 1943
Goodenough Island: July – November 1943
Kiriwina: November 1943 – June 1944
Tadji: June – August 1944
Noemfoor: August – November 1944
Moratai: November 1944 – June 1945
Tarakan: June – December 1945
Deniliquin: December – August 1946
Disbanded:
Richmond: March 1948 – March 1949
Schofields: March 1949 – October 1952
Fairbairn: October 1952 – April 1954
Richmond: April 1954 – March 1956
Disbanded:
Williamtown: January 1961 – December 1968
Disbanded: *December 2 1968*

---

31 Squadron formed at Wagga on August 14 1942 under the command of Squadron Leader C F Read. On receiving its first Bristol Beaufighter Mk.ICs early in September, the unit began training in the long range fighter and attack role before moving to Batchelor late in October. A further move to Coomalie Creek followed where operations commenced on November 17 when six Beaufighters attacked targets in Timor. Unfortunately, the formation was intercepted by Japanese fighters and one Beaufighter was lost.

On December 2 a very successful mission was flown when a formation of Beaufighters led by Flying Officer J E Dennett caught 40 Japanese fighters and bombers on the ground at Penfoei. Surprise was complete and 18 aircraft were destroyed for no loss. Operating at long range – Coomalie was 160kms south of Darwin – the Beaufighters continued their strikes against targets on Timor, and the Aru and Tanimbar Islands.

31 Squadron scored its first air-to-air victory on December 23 when Sergeant E J C Barnett shot down a Nate in a head on attack over Fuilora while, on February 23 1943, another very successful strike was made against Penfoei airfield when 12 aircraft were destroyed and 10 damaged despite determined fighter opposition and anti-aircraft fire which damaged one Beaufighter.

Japanese retaliatory raids were nowhere near as successful as evidenced on March 3 when enemy aircraft attacked Coomalie Creek damaging two Beaufighters and injuring two personnel.

Commencing in May, 31 Squadron expended considerable effort in destroying a Japanese float plane base at Tabufane in the Aru Islands. The first such operation was flown on May 6 when five Beaufighters strafed and destroyed nine enemy aircraft on the water. On June 4, four Beaufighters were engaged over Tabufane by nine floatplanes, three of which were shot down for no loss. Again, on June 12, seven floatplanes were destroyed on the water and a further two damaged. Successful attacks against this base continued despite increased anti-aircraft defences which damaged several Beaufighters. Eventually, the Japanese were unable to sustain their losses and evacuated Tabufane.

By Christmas 31 Squadron had destroyed 18 aircraft in the air and 49 on the ground – mostly at Tabufane and Penfoei. Several Beaufighters were lost achieving these results and as they were invariably hit at low level, their crews stood little chance of survival.

Operations continued at a high pitch throughout 1943. Later in the year, underwing bombs came into general use to supplement the Beaufighter's

31 Squadron Beaufighter.

already heavy armament of four 20mm cannons and four .50 calibre machine guns. Later still, devastating underwing rockets were utilised.

One of 31 Squadron's most successful pilots in this period was Squadron Leader R L Gordon. This officer destroyed several Japanese aircraft in air-to-air combat, including two Nicks – twin engined fighters similar in appearance to the Beaufighter – in a single mission. He was unfortunately killed on a nonoperational test flight in February 1944.

By 1944 worthwhile targets were becoming scarce in 31 Squadron's area of operations. Consequently, the Beaufighters flew at extreme range in search of their quarry, so far, in fact, that on many operations they had to fly at the most fuel efficient speed, even while in combat. To have increased power would have left insufficient fuel for the return flight to Australia.

31 Squadron began operations from Noemfoor and then Moratai in December with attacks against targets in the Celebes, Ambon, Ceram and the Halmaheras, having provided limited support to the Tarakan invasion.

After the war 31 Squadron conducted weather reconnaissance, dropped surrender leaflets to the Japanese and escorted single engined fighters on their long ferry flights back to Australia. 31 Squadron moved to Deniliquin in December and then to Williamtown where it disbanded on July 9 1946.

Formed: *August 14 1942*
Squadron Code: *EH*
Aircraft: *Beaufighter*
Locations: *Wagga: August – October 1942*
Batchelor: October – November 1942
Coomalie Creek: Nov 1942 – Dec 1944
Noemfoor: December 1944
Moratai: December 1944 – May 1945
Tarakan: May – August 1945
Moratai: August 1945 – December 1945
Deniliquin: December 1945 – March 1946
Williamtown: March 1946 – July 1946
Disbanded: *July 9 1946*

---

32 SQUADRON

32 Squadron was hastily formed at Port Moresby on February 21 1942 with Hudsons and personnel from 6, 23 and 24 Squadrons. Commanded by Wing Commander D Kingwell, the unit commenced reconnaissance operations the very day of its formation.

32 Squadron's first attack mission was flown on February 28 when targets at Gasmata were bombed. Reconnaissance and bomber operations continued at an intensive rate as the Hudson crews covered vast tracks of ocean searching for enemy shipping during a crucial phase of the Pacific war. From the start, Japanese fighters interfered with the unit's operations and Hudsons often returned to base damaged by gunfire with wounded and dead crewmen on board.

Port Moresby itself also came under regular air attack although 32 Squadron usually escaped without much damage. This was not the case however, on February 24, when ten Japanese bombers struck, demolishing much of the unit's camp, wrecking a Hudson and killing one airman.

An especially significant operation was flown on March 7 when Flying Officer Herme's crew located an 11 ship convoy headed towards Salamaua. Attacks against this convoy commenced once this crew's sighting was made known. One 8160 tonne vessel was hit and set on fire by another Hudson captained by Wing Commander Kingwell.

Pilot Officer P J E Pennycuick's crew located another convoy consisting of an aircraft carrier, a cruiser and several other ships on May 6 – information which proved to be of great value to Allied commanders during the Battle of the Coral Sea. A Hudson captained by Flight Lieutenant L W Manning made another important discovery on July 21 when Japanese troops were observed landing at Buna.

In the critical ground campaign now being fought in New Guinea, 32 Squadron, already heavily committed to reconnaissance and attack operations, began hazardous supply dropping missions to Australian troops. Although not especially suited to this task, the Hudsons were among the few aircraft available and the supplies they delivered had a direct bearing on the eventual success of the campaign.

During the battle for Milne Bay, Hudsons flew reconnaissance and bomber missions in support of the embattled Australian ground forces. On August 25 Flying Officer Williams located a number of invasion barges in the vicinity of Goodenough Island. Using this information, Kittyhawks attacked and destroyed these vessels. Afterwards, it was found that the destruction of

Upon reforming, 32 Squadron was equipped with HS.748s taken over from the School of Air Navigation. (Lance Higgerson)

the barges severely disrupted Japanese ship to shore transport and was a major factor in the Australian victory at Milne Bay.

After this critical battle had been won, 32 Squadron, which had been operating from Horn Island – with detachments at Port Moresby and Townsville – was withdrawn to Richmond and subsequently to Camden. From its new base, the unit conducted anti-submarine patrols, seaward searches and convoy escort missions. In January 1943 two attacks against Japanese submarines were made, however, in both cases the submarines escaped damage. Several other sightings and attacks were made later in the year.

32 Squadron received its first Australian built Bristol Beauforts in late June and began operating major detachments from Coffs Harbour and Bundaberg. Patrol operations continued with

less urgency until the end of the war. After moving to Lowood, 32 Squadron disbanded on November 30 1945.

Equipped with HS.748 turboprops, 32 Squadron reformed at East Sale on July 1 1989 and today operates in support of the School of Air Navigation and in the light transport role, mainly in southern Australia.

Formed: *February 21 1942*
Squadron Code: *JM*
Aircraft: *Hudson, Beaufort, HS.748*
Locations: *Port Moresby: Feb – April 1942*
Horn Island: April – September 1942
Richmond: September – November 1942
Camden: November 1942 – May 1944
Lowood: May 1944 – November 1945
Disbanded:
East Sale: July 1989 – present

---

33 SQUADRON

33 Squadron formed at Townsville on February 16 1942. As a transport squadron, the unit was initially equipped with impressed Short Empire flying boats and later a variety of lighter aircraft comprising de Havilland Dragon, Avro Anson, de Havilland Tiger Moth and Vultee Vigilant. With this assortment of makeshift transport aircraft operations were conducted throughout the Australian east coast. Possessing a longer range and greater payload, the Empires operated throughout Australia and sometimes further afield.

An unfortunate brush with the Japanese occurred on March 3 when an Empire captained by Flight Lieutenant Caldwell was caught on the water at Broome when nine Zero fighters attacked that port. The big flying boat was riddled

with cannon fire, burst into flames and sank. The crew escaped into a dinghy and then participated in the rescue of survivors from other flying boats which had suffered a similar fate as their own.

In January 1943 the unit moved to Port Moresby where its small aircraft flew intensively carrying freight and conducting communications tasks. Many freight runs to Myola and Kokoda were made – even the diminutive Tiger Moths were pressed into service and delivered 77 kilograms of cargo each trip!

33 Squadron began to re-equip with Dakotas from October 1943 and extended its operations over a much greater area. The first weeks of 1944 saw a move to Milne Bay affected while almost exactly one year later, in January 1945, a further move to Lae was made.

Former Qantas Empire flying boats were operated by 33 Squadron in the transport role.

After Japan's surrender, 33 Squadron evacuated liberated POWs from Singapore and then returned troops and equipment from the outlying island bases to Australia. The unit disbanded at Townsville on May 13 1946.

Over 30 years later on July 1 1983, 33 Squadron came into existence a second time when 33 Flight at Richmond was given squadron status. Equipped with Boeing 707s, the unit's role is that of strategic transport, VIP transport and air-to-air refuelling of the RAAF's fighter force. The latter capability, acquired in 1990, is crucial to the effective operation of the RAAF's fighter squadrons and allows the Hornets to project a defensive and offensive air combat capability a great distance from Australia's shores. 33 Squadron's operational flexibility is further enhanced by the Boeing 707's ability to operate in the combi role – that is, the aircraft can be configured to carry a mix of passengers and cargo during the same mission.

Aside from its VIP tasks, 33 Squadron has undertaken many important operations since reforming. Until the RAAF wound down its presence at Butterworth in the late 1980s, regular courier flights were made to Malaysia while, in 1989, the unit deployed Australian troops to Namibia where they participated in a major United Nations action.

Despite the age of its Boeing 707s, 33 Squadron, with its strategic transport and air-to-air refuelling capability, provides a vital and unique link in Australia's defence.

Formed: February 16 1942
Squadron Code: BT
Aircraft: Dragon, Anson, Empire, Tiger Moth, Vigilant, Ford 4-AT-E Trimotor, Miles M.3A, Catalina, Dakota, Boeing 707
Locations: Townsville: Feb 1942 – Jan 1943
Port Moresby: January 1943 – January 1944
Milne Bay: January 1944 – January 1945
Lae: January 1945 – March 1946
Townsville: March 1946 – May 1946
Disbanded:
Richmond: July 1983 – present

---

34 SQUADRON

On February 2 1942 34 Squadron formed at Darwin. Like other RAAF transport squadrons formed at this time, the unit, commanded by Flight Lieutenant J W Warwick, was not allocated aircraft commensurate with its role, and initially operated just two small de Havilland Dragons.

Ansons and Tiger Moths soon replaced the Dragons, one of which had been destroyed on March 3 during the Japanese attack on Wyndham. After operating from Batchelor, Hughes and Manbulloo, 34 Squadron disbanded at the latter location in December 1942.

Reforming at Parafield on January 1 1943, 34 Squadron was again initially equipped with a small allocation of biplane Dragons. Despite the inadequacy of these aircraft, transport operations were flown throughout South Australia and the Northern Territory.

34 Squadron finally began to receive more capable aircraft in the form of Douglas C-47

Dakotas in May 1943. The arrival of these robust and efficient transports allowed general freight and troop carrying operations to be extended throughout Australia, into New Guinea and later, much of South East Asia. Outside Australia supply dropping operations and medical evacuation flights were conducted. An Airspeed Oxford and a Douglas DC-2 were allocated to 34 Squadron later in the year.

BAC 111 VIP transport of 34 Squadron, with a Viscount and Dakota in the background.

In October 1944 a detachment of three aircraft began operations from Higgins, on Cape York Peninsula where flights into Tadji, Noemfoor, Hollandia and other forward locations were initiated. Other major detachments were later based at Townsville and Coomalie Creek. Operations were further extended when 34 Squadron moved to Moratai in February 1945. From here the Dakotas supported the invasion of Borneo and a 34 Squadron Dakota was, in fact, the first aircraft to land at Labuan's airfield after that island's seizure in June.

Following Japan's surrender, 34 Squadron evacuated Australian POWs from Singapore and then returned military personnel to Australia. As these tasks were completed, the unit commenced courier flights into Japan in support of the Allied occupation of that country. During February 1946 34 Squadron returned to Australia and disbanded at Richmond in June of the same year.

34 Squadron again reformed at Mallala on March 1 1948 when 2 Squadron was renumbered. Operating in the communications, VIP transport and range reconnaissance roles, the unit possessed a remarkably varied fleet of aircraft comprising Bristol Freighter, Prince, Viking, Auster, Dakota and Anson aircraft. Principally, 34 Squadron's operations supported the various activities undertaken at the Woomera rocket range although some transport tasks and civil aid flights were made. The unit disbanded at Mallala on October 28 1955.

A new phase began for 34 Squadron in July 1959 when 34 Flight at Fairbairn was given squadron status. Equipped with Convair Metropolitans and Dakotas, the unit's role was that of VIP transport.

New aircraft were received in October 1964 when Viscounts were delivered, while in April 1967 the first of two HS.748s arrived at Fairbairn. Two months later Dassault Mystère 20s were also allocated. Another addition to 34 Squadron's VIP fleet was the BAC 111 which arrived on January 19 1968. In September 1990 the first of five Dassault Falcon 900 medium range trijet transports was accepted, this type subsequently replacing all three previous types in service.

As a VIP squadron, the unit has carried members of the Royal Family, foreign dignitaries and heads of state, the Australian Governor General and politicians. VIP flights are conducted throughout Australia and sometimes overseas. With its modern Falcon aircraft and an outstanding record for safety and efficiency, 34 Squadron continues to provide a high standard of special transport for the Australian Government.

Formed: February 23 1942
Squadron Code: FD
Aircraft: Dragon, Anson, Tiger Moth, Dakota, Oxford, DC-2, Bristol Freighter, Prince, Viking, Auster, Metropolitan, Viscount, HS.748, Mystère 20, BAC 111, Falcon 900
Locations: Darwin: February – May 1942
Batchelor: May – July 1942
Hughes: July – August 1942
Manbulloo: August – December 1942
Disbanded:
Parafield: January 1943 – February 1945
Moratai: February 1945 – February 1946
Richmond: February – June 1946
Disbanded:
Mallala: March 1948 – October 1955
Disbanded:
Fairbairn: July 1959 – present

---

35 SQUADRON

Initially equipped with just two light aircraft – a Dragon and a Fox Moth – 35 Squadron formed at Pearce on March 11 1942.

The new Squadron was commanded by Flight Lieutenant P Burdeu and commenced freight and passenger missions immediately before moving to Marylands shortly afterwards.

To enhance its ability to conduct transport operations throughout Western Australia, 35 Squadron received additional aircraft throughout 1942 and into the following year. These included Moth Minors, Fairy Battles, Tiger Moths and single examples of the Anson, Dragon Rapide and Northrop Delta.

After returning to Pearce in the first week of August 1943, 35 Squadron received its first Dakotas. As the conglomeration of smaller aircraft was gradually replaced, the unit began to operate Australia wide.

In August 1944, 35 Squadron relocated to Guildford with a detachment and almost immediately commenced operations from Brisbane. A further detachment started flying from Higgins, on Cape York, in mid October. This detachment conducted courier runs to Aitape, New Guinea and to other locations as required.

As the Pacific war receded from Australia, 35 Squadron's aircraft moved to more northerly locations, the Brisbane detachment going to Townsville and a new detachment was established at Darwin. By now the Dakotas were ranging over much of the South West Pacific and often flew over enemy occupied territory. 35 Squadron's main party moved to Townsville between late January and March 1945 and in April

a detachment began operations from Moratai.

After Japan's surrender the Dakotas continued their usual transport tasks but ranged further afield throughout the Pacific. On August 25 Squadron Leader A Page's crew flew the first 35 Squadron Dakota flight to Singapore while in November Flight Lieutenant S Eddy's crew made the unit's first flight to Japan. In this period, Australian troops, including recently liberated POWs, were flown back to Australia. In early 1946, 35 Squadron supported the movement of three RAAF fighter squadrons and various support units to Japan where this 1700 man force was to form part of a military occupation force. This task was destined to be 35 Squadron's last major operation and the unit disbanded at Townsville on June 10 1946.

On June 1 1966, the RAAF Transport Flight – Vietnam (RTFV), at Vung Tau, South Vietnam, was retitled 35 Squadron. There were no frontlines in Vietnam and the rugged de Havilland Canada DHC-4A Caribou aircraft with which the unit was equipped could expect to come under fire at any time. Even 35 Squadron's home base at Vung Tau was subject to occasional mortar and rocket attacks.

35 Squadron operated cargo and passenger flights throughout South Vietnam. Paratrooping operations in support of the South Vietnamese Army were also a feature of 35 Squadron's role, as were occasional night flare dropping missions. Other tasks included medical evacuation and supply dropping.

RAAF Caribou, operating in a predominantly American air war, achieved some remarkable

Australian Caribou parked at Vung Tau.

results in Vietnam. While only flying 1.4% of the freight missions, the unit delivered no less than 7% of the total freight airlifted. This result so stunned the American command that it sent efficiency experts to monitor 35 Squadron's daily activities. Whatever their findings, it remains history that the United States Caribou squadrons never managed to operate as effectively as their Australian counterparts.

35 Squadron was required to land on many remote and difficult landing grounds, usually in support of isolated Special Forces garrisons. The poor state of one of these was responsible for serious damage being sustained by one of the unit's Caribou while landing at Ba To Special Forces camp on August 16 1966. Not prepared to abandon a retrievable aircraft, a ground staff team from Vung Tau was flown in and, working under appalling conditions, repaired the Caribou which was successfully flown out. While at Ba To the Australian ground staff had to endure an enemy ground assault on the camp which was successfully beaten off. In July 1967 another damaged Caribou was recovered in similar circumstances.

The Caribou, at times, were forced to operate to very low level in dangerous tropical storms and this contributed to several aircraft being hit by small arms fire. As a result of this fire a number of aircrew and their passengers were wounded – two aircrew being especially lucky to survive after being hit in the head.

The next major incident involving the enemy and a Caribou occurred on January 19 1969 when an aircraft captained by Flight Lieutenant T Thompson was landing at Katum Special Forces Camp. Just after touchdown the Caribou came under accurate mortar fire. One round landed within eight metres of the aircraft, holing it with shrapnel in over a hundred places. The hydraulics, brakes and flaps were all damaged and both main wheel tyres were punctured. Flight Lieutenant Thompson was wounded by shrapnel in the leg and his copilot, Flying Officer R McGregor injured by flying perspex fragments from their shattered windscreen.

To have remained on the ground would have resulted in the destruction of the aircraft so an immediate takeoff was initiated. Meanwhile, the two loadmasters, mindful of 35 Squadron's proud boast to always deliver their cargo, pushed Katum's consignment of stores out onto the runway as the aircraft rolled down the strip. The badly damaged Caribou was later repaired at Vung Tau.

Not so fortunate were the crew of Pilot Officer B Milne's Caribou who were mortared while delivering aviation fuel to That Son on March 29 1970. Hit in the wing by one round, the Caribou burst into flames and was completely destroyed.

The airmen escaped without injury and were later returned to Vung Tau.

With Australia's involvement in Vietnam decreasing, three of 35 Squadron's Caribou and 44 personnel returned to Australia in June 1971. Operations continued with the remaining aircraft and by Christmas 1971, 35 Squadron was the last remaining RAAF squadron in Vietnam. Time was rapidly running out for the unit, however, and on February 13 operations ceased and the remaining Caribou left for Australia seven days later.

While in Vietnam, 35 Squadron, and its predecessor, the RTFV set new standards for Caribou operations and maintenance and made a significant contribution in support of the Allied ground forces during the conflict.

Four years after arriving back at Richmond, 35 Squadron moved to Townsville where, early in 1977, four Bell UH-1 Iroquois joined its three Caribou. 35 Squadron thus became the only RAAF squadron to operate a combination of fixed and rotary wing aircraft. With its mixed fleet of aircraft, 35 Squadron undertook army tactical support tasks and also various activities in support of the civilian population, such as search and rescue operations, medical evacuations and flood relief work.

In November 1986, 35 Squadron began gunship operations in support of the Australian Army. One year later, the unit received a large increase in strength when it took over all of 9 Squadron's Iroquois pending that unit's re-equipment with Black Hawks. 35 Squadron continued its mixed rotary/fixed wing operations until December 5 1989 when the Iroquois were transferred to Army control.

Currently 35 Squadron operates eight Caribou, two of which are permanently detached at Darwin. The unit remains heavily committed to army support work in northern Australia.

Formed: *March 11 1942*
Squadron Code: *BK*
Aircraft: *Fox Moth, Dragon, Moth Minor, Battle, Tiger Moth, Anson, Dragon Rapide, Delta, Dakota, Caribou, Iroquois*
Locations: *Pearce: March – April 1942*
Marylands: April 1942 – August 1943
Pearce: August 1943 – August 1944
Guildford: August 1944 – March 1945
Townsville: March 1945 – June 1946
Disbanded:
Vung Tau: June 1966 – February 1972
Richmond: February 1972 – July 1976
Townsville: July 1976 – present

---

36 SQUADRON

Formed at Laverton on March 11 1942, 36 Squadron, under the command of Flight Lieutenant W P Heath, was initially equipped with just one Douglas DC-2 and 26 personnel.

Transport operations commenced almost immediately despite the newly formed unit's inadequate facilities – squadron offices, for instance, were converted packing cases – lack of personnel and aircraft spares. Gradually the unit was increased in size until it operated a conglomeration of aircraft including six DC-2s, two de Havilland DH.86s and single examples of the Dragon Rapide, Fairchild 24, Tiger Moth, Stinson L-5 Sentinel, Ford 5-AT-C and Junkers G31 Tri Motor.

In mid July the unit moved to Essendon where facilities were infinitely better. Operations continued at great intensity throughout Australia particularly into Darwin and later New Guinea. By December 1942, 36 Squadron was operating from Townsville where it received its first Douglas Dakotas. The conglomeration of aircraft previously operated was gradually replaced until the unit was largely re-equipped with these aircraft, although a detachment of DC-2s was maintained at Richmond for parachute training duties.

Food and equipment was continually flown to New Guinea and the first of several detachments to that combat zone commenced in 1943. These detachments conducted supply dropping, troop transport and freight runs often over difficult terrain and in dangerous weather conditions. As the Allies forced the enemy back towards the Japanese home islands 36 Squadron's operations were extended over an ever increasing area – often over enemy occupied territory.

After the Japanese surrender 36 Squadron dropped supplies to Allied POWs and later commenced POW evacuation flights from Singapore, Thailand and other areas. From March 1946 a six aircraft detachment at Moratai began courier runs to Japan in support of the Australian component of the British Commonwealth Occupation Force.

36 Squadron had moved from Townsville to Schofields by mid August 1946 but continued courier flights to Japan and indeed throughout the South Pacific. The Japan courier at that time was the longest air transport route in the world using twin engined aircraft. In August 1948 half of the unit's aircrew were sent to Europe to participate in the Berlin Airlift and for many months 36 Squadron was virtually combined with 38 Squadron due to a lack of personnel.

36 Squadron moved to Richmond in June 1949. Aside from maintaining detachments at various locations – in particular Townsville – and flying cargo and personnel, the Dakotas regularly conducted civil aid tasks including flood relief and bushfire spotter patrols.

On March 8 1953, 36 Squadron ceased to exist at Richmond (formally disbanding on May 26 1953). On March 10, 30 (Transport) Unit at Iwakuni, Japan was renamed 36 Squadron. From Iwakuni, 36 Squadron supported the United Nations forces in Korea by carrying freight to and from Japan, evacuating casualties and POWs and providing a VIP transport capability for the United Nations Command. After the armistice in July 1953, 36 Squadron remained in Japan supporting a continued United Nations presence in Korea.

A 36 Squadron Dakota sheds its cargo during the Korean War. (RAAF Museum)

A C-130H Hercules of 36 Squadron drops supplies. (RAAF Museum)

Reduced to flight status, the unit moved to Fairbairn on March 18 1955, where it once again regained squadron status on May 1 1955. Routine transport operations continued until July 1958 when most of 36 Squadron's personnel deployed to the United States to convert onto the Lockheed C-130A Hercules. Once their training had been completed, these personnel ferried their new aircraft to Richmond, the unit's new base, between December 1958 and March 1959.

With its new aircraft, 36 Squadron was well placed to conduct strategic and tactical transport operations throughout Australia and South East Asia. An early task which tested the capability of 36 Squadron was the deployment and maintenance of 79 Squadron to Ubon, Thailand. The Hercules also proved highly suited to a variety of civil aid tasks such as fodder drops during floods, air sea rescue work and medical evacuations.

With the escalating commitment of Australian forces in Vietnam during the mid 1960s, 36 Squadron found itself operating a regular courier service to and from that country, carrying troops and equipment. Prior to the reformation of 37 Squadron with the later C-130E Hercules, 36 Squadron also conducted medical evacuation flights from Vietnam. It was a proud boast that not one patient, no matter how severely wounded, died during these flights.

Since the end of the Vietnam War, 36 Squadron, which re-equipped with C-130H Hercules in 1978, has continued to support Australian Defence Force operations within Australia and indeed throughout the world. The unit has also supported the United Nations in various countries including Cambodia and Somalia. 36 Squadron also played a pivotal role in the relief and subsequent evacuation of Darwin after that city was destroyed by Cyclone Tracey in December 1974. One unusual activity conducted on behalf of the Australian Government involves the support of scientific expeditions in the Australian Antarctic Territory. This support has involved temporary operation of the Hercules from the ice cap itself.

36 Squadron continues to operate from Richmond and in the next few years will likely be equipped with new aircraft – most likely the latest version of the ever faithful C-130 Hercules.

Formed: *March 11 1942*
Squadron Code: *RE*
Aircraft: *DC-2, DH.86, Dragon Rapide, Fairchild 24, Tiger Moth, Stinson L-5, Ford Trimotor, Dakota, Hercules*
Locations: *Laverton: March – July 1942*
Essendon: July – December 1942
Aitkenvale: December 1942 – February 1944
Garbutt: February 1944 – August 1946
Schofields: August 1946 – June 1949
Richmond: June 1949 – March 1953
Iwakuni: March 1953 – March 1955
Disbanded:
Fairbairn: May 1955 – October 1958
Richmond: October 1958 – present

---

37 SQUADRON

37 Squadron formed at Laverton on July 15 1943. The new transport unit, commanded by Squadron Leader N Hemsworth, was equipped with Lockheed Lodestar twin engined transports and, for a short period, a single Northrop Delta.

Transport operations commenced on September 21 and, as the unit received more Lodestars, regular courier runs were conducted to Western Australia, Tasmania, New Guinea and the Northern Territory. Operations into the New Guinea area were extended to Aitape in July – a distance of some 11,300 kilometres return to Laverton. Later, extensions encompassed Noemfoor, Moratai and other island bases.

37 Squadron moved to Essendon in September 1944, where in February 1945 it began to re-equip with Dakotas. After Japan's surrender the Dakotas evacuated Australian POWs from Singapore and returned Australian troops and equipment from island bases to Australia. From February 1946, 37 Squadron supported the deployment and maintenance of the British Commonwealth Occupation Force in Japan. Despite the arduous nature of the Japan courier flying routine, transport operations, throughout Australia and New Guinea continued in parallel. During July 1946, 37 Squadron moved to Schofields where it disbanded a year and a half later on February 24 1948.

After operating Lodestars 37 Squadron was later allocated Dakotas, a type which stayed in service for the remainder of World War II.

As a direct result of Australia's increasing involvement in South East Asia, the Australian Government ordered 12 C-130E Hercules to strengthen the RAAF's transport fleet in February 1965. On February 21 1966, 37 Squadron reformed at Richmond to operate the new aircraft and received its first Hercules in August with deliveries being completed by January 1967. Even before being fully equipped, 37 Squadron found itself flying long range transport missions in support of the Australian force in Vietnam. Troops and equipment were carried and, specially rigged Hercules, carried aero medical evacuation teams returning casualties to Australia.

Since the end of the Vietnam War 37 Squadron has continued to support the Australian Defence Force, both within Australia and abroad. The Hercules have regular ports of call in Malaysia,

A 37 Squadron C-130E Hercules with "Hawkesbury 200" markings. (Doug MacKay)

Singapore and New Zealand and at various times have operated in many other countries. Civil aid and humanitarian tasks have also played a major part in the unit's operation. In December 1974, 37 Squadron operated intensively, flying in supplies and evacuating people from cyclone ravaged Darwin.

The useful Hercules have been widely used for civil tasks and have proved particularly suited to fodder dropping during floods which periodically effect parts of Australia. For nearly four months from August 25 1989, 37 Squadron and the rest of the RAAF's transport force, conducted large scale airline style flying of civilian passengers around Australia during a major industrial dispute involving the nation's airline pilots. This task was faultlessly executed and proved to be a great credit to the RAAF.

Although 37 Squadron's Hercules are now badly in need of replacement (the squadron will re-equip later this decade with C-130Js), the unit continues to provide the Australian Defence Force and the Australian community with a high standard of airlift support.

Formed: *July 15 1943*
Squadron Code: *OM*
Aircraft: *Lodestar, Delta, Dakota, Hercules*
Locations: *Laverton: July 1943 – Sept 1944*
Essendon: September 1944 – July 1946
Schofields: July 1946 – February 1948
Disbanded:
Richmond: February 1967 – present

---

38 SQUADRON

Formed at Richmond on September 15 1943, 38 Squadron was initially equipped with Lockheed Hudsons and operated in the transport role. Eight months after formation, the Hudsons were replaced by Douglas Dakotas and with these larger aircraft operations increased in effectiveness and extended into New Guinea and other localities in the South West Pacific. In the forward areas, low level supply dropping missions, despite hazardous flying conditions, were conducted in support of Australian troops. In this role the ever reliable Dakotas became known as "Biscuit Bombers".

38 Squadron moved to Archerfield in December 1944, this move allowing it to operate closer to Allied forces in New Guinea. From July 1945, a detachment began flying from Moratai and after the Japanese surrender the unit evacuated liberated POWs and returned personnel and equipment to Australia.

In 1946, with its repatriation work completed, 38 Squadron moved to Schofields and two years later, to Richmond. From these airfields routine transport operations and occasional civil aid tasks were undertaken. During this period 38 Squadron, in turn with 36 Squadron, flew the Japan courier run – a thrice weekly service in support of the Australian component of the British Commonwealth Occupation Force. This demanding journey involved some 20,000 kilometres of travel and resulted in crews being away from Richmond for several days during each courier.

From late 1948 a large portion of 38 Squadron's aircrew strength was attached to the British RAF in Europe to fly Dakotas during the Berlin Airlift. The loss of these crews severely disrupted operations until replacements arrived. In June 1950 further upheaval occurred with the transfer of the whole unit to Changi, Singapore, where it was placed under the operational control of the RAF for service against Communist terrorist (CT) forces in Malaya.

After a short period of familiarisation flying and survival training, 38 Squadron flew its first operational missions. Tasks included supply dropping, general transport, casualty evacuation and VIP transport. Air sea rescue work, naval co-operation and leaflet dropping activities were also occasionally carried out. 38 Squadron's operations were not confined solely to Malaya and in fact extended as far afield as Ceylon (Sri Lanka), Korea, the Philippines and Japan.

In November 1950, 38 Squadron's complement of Dakotas was cut in half – from eight to four aircraft – due to pressing RAAF needs for this type of aircraft in Korea. The four Dakotas flew directly to Korea with air and ground crews and immediately began operations in support of United Nation forces. The remaining aircraft, comprising a much reduced 38 Squadron, continued their operations in support of British Commonwealth forces until December 1952 when the unit returned to Richmond.

After returning to Australia the Dakotas soldiered on for many years. Some unusual tasks ensued in this period as 38 Squadron was occasionally called on to support the work of the CSIRO during rain making experiments.

38 Squadron prepared for the long awaited replacement of its Dakotas with the de Havilland Canada DHC-4A Caribou in early 1964. The first of these new aircraft – which brought to the RAAF a robust aircraft with a remarkable short field takeoff and landing capability – arrived at Richmond on April 22 1964. The unit's full re-equipment, however, was hampered by the diversion of some aircraft to Vietnam. Later deliveries bought 38 Squadron up to strength and the unit quickly became proficient in its allotted tactical transport role.

During 1965, 38 Squadron began to operate a detachment from Port Moresby, where it supported both military activities and the civilian population for the next ten years. Flying conditions were extremely demanding and the detachment proved to be of great benefit in honing the Caribou crews' flying skills.

Another permanent detachment commenced in March 1975 when a white painted Caribou deployed to Rawlpindi in Pakistan to support the United Nations observer group monitoring the ceasefire between Pakistan and India. The Caribou also flew from Srinigar on the Indian side of the border and operated in mountainous terrain utilising small and unimproved landing grounds to supply isolated United Nations outposts. Having completed over 2000 accident free flying hours, the detachment returned to Australia in January 1979.

In August 1975, 38 Squadron also supplied an aircraft to support the Red Cross in Timor. The Caribou, operating from Darwin, ferried Red Cross personnel and medical supplies to and from Timor. During one flight on September 4 the aircraft was hijacked at gunpoint by a Timorese soldier and forced to fly back to Darwin where the offender was arrested without further incident.

With its easy access rear loading door and the ability to operate from unimproved landing strips, the Caribou has proved extremely useful during civil disasters. 38 Squadron has repeatedly participated in flood relief operations, including fodder drops to stranded cattle, searches for missing boats, bushfire patrols and a variety of other tasks. Some of the more unusual

Today, 38 Squadron continues to operate its faithful Caribou although the need to find a replacement is becoming urgent.

activities undertaken include fisheries surveillance and grasshopper plague eradication.

On the military side, the unit engages in a number of different activities including freight missions, paratrooping and the delivery of stores into small clearings using the LAPES low altitude parachute extraction system.

38 Squadron moved to Amberley in December 1992 where it continues to operate the Caribou in support of the Australian Army despite increasing difficulty in obtaining engine spares. For several years a replacement aircraft has been mooted but it is certain that 38 Squadron will continue to operate the type for some years to come.

Formed: September 15 1943
Squadron Code: PK
Aircraft: Hudson, Dakota, Caribou
Locations: Richmond: Sept 1943 – Dec 1944
Archerfield: December 1944 – August 1946
Schofields: August 1946 – July 1949
Richmond: July 1949 – June 1950
Changi: June 1950 – December 1952
Richmond: December 1952 – December 1992
Amberley: December 1992 – present

40 SQUADRON

40 Squadron formed at Townsville under the command of Wing Commander V A Hodgekinson on March 31 1944.

Allocated six Shorts Sunderland III flying boats, the new unit operated in a transport role. Despite a severe shortage of personnel the Sunderland crews immediately commenced operations and flew almost exclusively between Townsville and New Guinea. By July 40 Squadron had relocated to Port Moresby.

The RAAF's first attempted hijacking occurred at Townsville on September 14 1944 when Flight Lieutenant F V Mangers crew and passengers were held up at gunpoint by an American soldier not wishing to proceed to New Guinea. Eventually he was disarmed and arrested before anyone was hurt in the incident.

Occasional air sea rescue tasks were also conducted by 40 Squadron either while the Sunderlands were on normal transport flights or as special operations. During one of these on March 30 1945, Flight Lieutenant C B Hugall's crew located survivors from a crashed Dakota. After establishing their position for surface craft the Sunderland crew dropped a dinghy and medical supplies to the survivors.

By mid 1945 four Martin Mariner flying boats had been taken on charge and these were operated alongside the Sunderlands. With the cessation of hostilities

40 Squadron repatriated Australian personnel to the mainland prior to moving to Rathmines in March 1946. At this time, Catalinas replaced the Mariners, however, operations were rapidly scaled down and the unit officially disbanded on 19 June 1946.

Formed: *March 31 1944*
Squadron Code: *HF*
Aircraft: *Sunderland, Mariner, Catalina*
Locations: *Townsville: March – July 1944*
Port Moresby: July 1944 – March 1946
Rathmines: March – June 1946
Disbanded: *19 June 1946*

A Mariner flying boat of 40 Squadron. These robust transports were capable of taking off on one engine with a heavy load of cargo.

41 SQUADRON

Equipped with Short Empire S.23 flying boats, 41 Squadron formed at Townsville under the command of Squadron Leader J M Hampshire on August 21 1942. Carrying freight and passengers, the Empires flew their first transport operations that same day.

In their standard transport runs to New Guinea, the flying boats usually carried 1906 kilograms of freight, however, this could be increased to 2270 kilograms if necessary.

June 1943 saw the first of six ex Dutch Dornier Do 24K flying boats allotted to 41 Squadron as replacements for the Empires. Although impressive in appearance, these three engined German built aircraft could only carry 908 kilograms of freight and were in very poor mechanical condition. Consequently serviceability was low despite unceasing efforts by the ground staff. Apart from its transport activities 41 Squadron also carried out regular air sea rescue tasks in which over 150 personnel were recovered from the sea.

In February 1944, high performance Martin PBM-3R Mariner flying boats arrived to supplement the Dorniers. The new aircraft with their greater payload and performance quickly became the preferred aircraft, especially on the longer flights to Noumea, Espirito Santo and other island ports of call. Many of these flights originated from Cairns where 41 Squadron had moved in June 1944. However, 41 Squadron did not long survive Japan's surrender, disbanding on September 27 1945.

A Dornier Do 24K flying boat of 41 Squadron lies at anchor. The aircraft's impressive appearance was belayed by their poor mechanical condition and limited cargo capacity.

Formed: *August 21 1942*
Squadron Code: *DQ*
Aircraft: *Empire, Dornier Do 24K, Mariner*
Locations: *Townsville: Aug 1942 – May 1944*
Rathmines: May – June 1944
Cairns: June 1944 – September 1945
Disbanded: *September 27 1945*

42 SQUADRON

Equipped with Catalinas, 42 Squadron formed at Darwin on June 1 1944 under the command of Flight Lieutenant A M McMullin. The following month, the new flying boat squadron moved to Melville Bay where it became operational.

42 Squadron flew its first operational mission on August 27 when a shipping search was conducted off Timor. No sightings were made but Dili was bombed before the flying boat returned to Darwin. For a short time 42 Squadron carried out a variety of tasks including patrol and shipping escort work prior to being almost exclusively assigned to the RAAF's highly successful mining campaign.

In September 43 Squadron lost its first Catalina when, during a mine laying operation, Flight Lieutenant J M Cane's crew were forced down on the open sea with engine trouble. The aircraft floated all night and the anxious crew were rescued next day by another Catalina. Before taking off, the damaged flying boat was destroyed by machine gun fire to prevent its capture. October saw an intensive number of mine laying missions flown around Makasser in which several aircraft were hit and one lost to anti-aircraft fire.

42 Squadron's mine laying operations continued until the end of the war after which they were substituted by evacuation flights for released POWs and other personnel. A number of reconnaissance flights over Japanese occupied territory were also carried out. 42 Squadron disbanded at Melville Bay on November 30 1945.

Formed: *June 1 1944*
Squadron Code: *RK*
Aircraft: *Catalina*
Locations: *Darwin: June – July 1944*
Melville Bay: July – November 1945
Disbanded: *November 30 1945*

Ground crew pull a Catalina on its beaching gear.

43 SQUADRON

Commanded by Flight Lieutenant C Thompson, 43 Squadron formed at Bowen on May 1 1943. The new unit was equipped with Catalina flying boats and allocated a general reconnaissance role. However, as with all the RAAF's Catalina squadrons, it engaged in a variety of activities which would later encompass mining, bombing, air sea rescue and supply dropping.

Moving to Karumba in August, 43 Squadron flew its first operational mission on September 8 when four Catalinas participated in an attack against Ambon. This raid, in which large fires were started, was the first in a succession against this target.

Mining missions, mainly in Netherlands East Indies waters, were also undertaken and, although immediate results could not be seen, it was later found that this campaign had a crippling effect on Japanese shipping movements. Consequently these operations were considerably stepped up. 43 Squadron's first rescue was effected on January 19 1944 when four survivors from an American bomber were rescued from the open sea.

In April 43 Squadron moved to Darwin where it was assigned almost exclusively to mine laying. From Darwin, operations commenced on the night of April 17 when four flying boats, staging through Yampi Sound, mined Balikpapan harbour. The Catalinas met heavy anti-aircraft fire and one, captained by Flight Sergeant D Abby, was shot down. Despite this, and later losses, 43 Squadron operated intensively sowing mines right up until the end of the war.

On January 15 1945 Flight Lieutenant B Ortlepp's crew landed on the open sea in a heavy swell to rescue the crew of another Catalina who had force landed due to engine trouble. Despite the state of the sea the survivors were bought on board and flown back to Darwin. Before taking off, the damaged flying boat was destroyed by machine gun fire. This was the second rescue of this type conducted by 43 Squadron as, during the previous October, a similar feat was performed by Flying Officer A A Etienne's crew.

On the night of April 5/6, three Catalinas shadowed a Japanese convoy comprising the cruiser *Isuzu* and four smaller ships north of Darwin. Before dawn the flying boats had to leave the area due to the threat of fighter interception, however, the vessels were later successfully attacked by aircraft and submarines.

After the cessation of hostilities 43 Squadron carried out routine transport missions until moving to Rathmines in November where it disbanded on April 10 1946.

Formed: *May 1 1943*
Squadron Code: *OX*
Aircraft: *Catalina*
Locations: *Bowen: May – August 1943*
Karumba: August 1943 – April 1944
Darwin: April 1944 – November 1945
Rathmines: November 1945 – April 1946
Disbanded: *April 10 1946*

One of 43 Squadron's Black Cats.

60 SQUADRON

Under the command of Squadron Leader B R Pelly, 60 Squadron formed at Wagga on January 1 1942. Equipped with CAC Wirraways, the unit's formation was directly attributable to Japan's spectacular successes in the first months of the Pacific war and the consequent threat of Australia being invaded.

Formation flying, gunnery, and dive bombing practice were undertaken as a matter of urgency both at Wagga and after the unit moved to Cootamundra in February.

60 Squadron's existence was to be a short one, however, as its aircraft and personnel were urgently required by training units. Consequently, it disbanded on April 3 1942.

Formed: *January 1 1942*
Squadron Code: *EY*
Aircraft: *Wirraway*
Locations: *Wagga: January – February 1942*
Cootamundra: February – April 1942
Disbanded: *April 3 1942*

66 SQUADRON

66 Squadron formed at Bundaberg on May 20 1943. The formation of the new unit, commanded by Squadron Leader R F Wiley, was an emergency measure in response to Japanese submarine activity off the east coast of Australia. 66 Squadron's Avro Anson Mk.Is and personnel were all drawn from Bundaberg based 8 Service Flying Training School.

Anti-submarine patrols and convoy escort missions commenced immediately, however, little was sighted on these flights.

With a reduction of enemy submarine activity apparent by the end of 1943, the requirement for anti-submarine aircraft lessened and, as 66 Squadron's personnel and Ansons were urgently required for training duties, the unit disbanded on January 6 1944.

Formed: *May 20 1943*
Squadron Code: *JN*
Aircraft: *Anson*
Locations: *Bundaberg: May 1943 – Jan 1944*
Disbanded: *January 6 1944*

67 SQUADRON

Equipped with Anson Mk.Is, 67 Squadron formed at Laverton on January 6 1943 under the command of Flying Officer W J Canterbury. Operating from its home base as well as Mallacoota, Yanakie, Bairnsdale and Warnambool, the unit conducted anti-submarine patrols and convoy escort missions around southern Australia.

Apart from this generally uneventful work, 67 Squadron also carried out searches for missing aircraft – both civil and military – and overdue vessels. Additionally, naval and army co-operation exercises were also conducted

Between April 21 and 24 1945 an Anson carrying a member of the Council for Scientific and Industrial Research, performed what was perhaps 67 Squadron's most unusual task when it was used to conduct a Pelagic Fish Survey between Sydney and Ceduna. 67 Squadron disbanded at Laverton on November 10 1945.

Formed: *January 6 1943*
Squadron Code: *MK*
Aircraft: *Anson*
Locations: *Laverton: Jan 1943 – Nov 1945*
Disbanded: *November 10 1945*

67 Squadron Ansons at Laverton in 1945.

71 SQUADRON

To combat Japanese submarine attacks off Australia's east coast, 71 Squadron, equipped with Anson Mk.Is, formed at Lowood on January 26 1943. Squadron Leader P L B Gibson was appointed to command the new unit.

Operating detached Anson flights from Amberley, Bundaberg, Richmond and Coffs Harbour, anti-submarine patrols and convoy escort missions were flown. These resulted in a few submarine sightings and one unsuccessful attack on March 17 1943. Occasional air sea rescue tasks, as well as army and naval co-operation exercises, were carried out before the unit disbanded at Coffs Harbour on August 28 1944.

Formed: *January 26 1943*
Squadron Code: *PP*
Aircraft: *Anson*
Locations: *Lowood: Jan 1943 – March 1944*
Coffs Harbour: March – August 1944
Disbanded: *August 28 1944*

73 SQUADRON

Commanded by Flying Officer K C Berry, 73 Squadron formed at Cootamundra on May 8 1942. Equipped with Anson Mk.Is, the unit conducted anti-submarine patrols and convoy escort missions off Australia's east coast.

Detachments operated from Nowra, Richmond, Moruya and later Coffs Harbour and Camden, however, submarines were rarely sighted. By 1944, Japanese submarine menace was almost non existent and 73 Squadron disbanded at Nowra on September 9 1944.

Formed: *May 8 1942*
Squadron Code: *NJ*
Aircraft: *Anson*
Locations: *Cootamundra: May – Dec 1942*
Nowra: December 1942 – September 1944
Disbanded: *September 9 1944*

75 SQUADRON

One of the RAAF's most famous units, 75 Squadron formed at Townsville on March 4 1942.

Equipped with American built Curtiss P-40E Kittyhawk fighters, the unit, commanded by Squadron Leader P Jeffrey, came into existence at a time when most of New Guinea had fallen to the Japanese and it appeared that Australia itself was in imminent danger of suffering the same fate.

Due to this critical situation, only nine days were available to train 75 Squadron to an operational standard – surely one of the shortest periods in the history of air warfare – after which the Kittyhawks flew to Port Moresby on March 21.

Port Moresby had been under constant air attack for some time and, as the fighters prepared to land, anti-aircraft gunners anticipating yet another air raid, opened fire and damaged three aircraft. Just hours after arriving a Japanese bomber on a reconnaissance mission appeared overhead. Two Kittyhawks, flown by Flying Officers B M Cox and W L Wackett were scrambled and shot the aircraft down, much to the jubilation of the watching Australian garrison.

Next morning, nine Kittyhawks took the offensive to the Japanese by attacking Lae airfield. A dozen enemy aircraft were destroyed on the ground and a further five damaged. As they completed their attack, the Kittyhawks were intercepted by three Zeros. In the ensuing combat two Kittyhawks were shot down (the pilot of one – Flying Officer W L Wackett – survived a ditching in the sea and subsequently returned to Port Moresby) while two Zeros were damaged.

After this most successful attack, 75 Squadron – the only Allied fighter squadron then in the whole of New Guinea – flew daily in the defence of Port Moresby. Always heavily outnumbered and operating under primitive conditions, the unit nonetheless extracted a continuous toll on the Japanese. After an abortive interception of an incoming raid on March 23, the next day saw three enemy aircraft fall to the Australian pilots. Opposing one raid single handedly, Flight Lieutenant L Jackson took off and intercepted 18 bombers and three fighters. Despite the attention of the fighter escort the Australian succeeded in shooting down one of the bombers.

Combats continued almost on a daily basis and the heavily outnumbered Australian squadron was quickly reduced in strength. An early attempt to withdraw was, however, rejected with the full support of the unit's personnel, and with replacement aircraft and pilots, 75 Squadron fought on. Unfortunately, the appalling conditions under which all personnel laboured was taking a heavy toll on the men's health. Many fell prey to malaria and other tropical illnesses while extremely long working hours and strain created by enemy attacks and poor living conditions further exacerbated health problems.

On April 28, 75 Squadron engaged in its last major combat from Port Moresby when its five remaining serviceable Kittyhawks encountered a large force of bombers and fighters whilst headed towards Lae. Despite being outnumbered, the aggressive Australian pilots turned to the attack and both Squadron Leader J Jackson and Flight Lieutenant Cox destroyed a Zero before they were themselves shot down and killed. A third Kittyhawk was damaged.

A few days later, a battle weary 75 Squadron, with just one serviceable Kittyhawk left, was relieved and returned to Australia to retrain and re-equip. While engaged in its epic 44 days of combat, 75 Squadron destroyed 17 Japanese air-craft in air-to-air combat, probably destroyed four and damaged a further 29. Additionally, 17 aircraft were destroyed on the ground and another 15 damaged. Losses had been heavy – 12 pilots had been killed and 22 aircraft destroyed. Notwithstanding its own heavy losses, 75 Squadron had conducted a heroic and able defence of Port Moresby in what was perhaps the most crucial period of the New Guinea campaign.

75 Squadron's well deserved rest was cut short, however, when in late July, it returned to New Guinea. Commencing operations from Gurney strip at Milne Bay on July 25, 75 Squadron joined 76 Squadron at that base – just in time to become embroiled in another of the south west Pacific campaign's most critical battles.

The newly arrived squadron did not have long to wait before clashing with the enemy when, on August 11, 12 Zeros approached Milne Bay. These aircraft were intercepted by 22 Kittyhawks from both squadrons. Despite being outnumbered the Japanese inflicted losses on 76 Squadron and shot down 75 Squadron's Flying Officer M E Sheldon and Warrant Officer F P Shelley. In return Flying Officer G Atherton probably destroyed two Zeros while 76 Squadron shot down one enemy aircraft.

On August 24 nine Kittyhawks were sent to attack Japanese barges located on Goodenough Island. Destroying these vessels proved an easy task and the Australian pilots could not have realised that they had just conducted the first, and most decisive attack, in the battle for Milne Bay. Later that day, a Japanese invasion force was located steaming towards Milne Bay. The Kittyhawks had been modified to carry bombs and, prior to darkness falling, two strikes were launched against the approaching ships. These met with some success and one naval vessel, damaged by the bombing, lost way and did not proceed after the second attack.

Despite this setback, most of the convoy entered Milne Bay late that night, disembarked their troops and departed before dawn. At first light, the Kittyhawks began shuttle attacks against landing barges, stores and troops. These initial strikes were exceedingly effective and caused considerable loss to the Japanese.

Working in appalling conditions and torrential rain the Australian ground staff refuelled and rearmed their aircraft as quickly as possible and the Kittyhawks continued their attacks over the succeeding days. These operations were usually in support of the hard pressed Australian

Flying Officer T A Jacklin who was hit by anti-aircraft fire returned to Noemfoor in this badly damaged Kittyhawk which had less than 75% of its wing surface intact. (RAAF Museum)

During an armament camp in the Mediterranean, this 75 Squadron pilot and airmen check a Vampire's weapon load. (RAAF Museum)

troops who were slowly but inexorably being pushed back towards the RAAF airstrips – the Japanese objective. These attacks, like those made previously, were very effective and it has been recorded that so close was the enemy that the Kittyhawk's guns were chattering scarcely before their undercarriages retracted on takeoff.

In an ineffectual attempt to counter the damaging air attacks being mounted against their own troops, eight Japanese dive bombers and 12 Zeros attacked Gurney Strip on the morning of August 27. Six 75 Squadron aircraft intercepted and between them Flying Officers B D Watson and P B Jones shot down a dive bomber, probably destroyed a second and damaged a third. One Kittyhawk, flown by Flight Sergeant S Munro was lost. For their part, the Japanese aircraft failed to achieve any appreciable result and the attacks against their ground troops continued uninterrupted. In another combat later that day Squadron Leader L Jackson and Flight Sergeant R G Riddel bounced two Zeros, destroying both.

By now, so close were the Japanese that, on the afternoon of August 28, the Kittyhawks were flown to the relative safety of Port Moresby for the night to prevent their possible capture. Next morning, however, they were back supporting the staunch Australian ground forces who continued to contest every metre of ground. By September 1 it was becoming apparent that the Japanese were losing the battle and pressure on the Australian troops gradually decreased. Nightly, Japanese ships were entering Milne Bay and, aside from bombarding shore targets, were embarking what troops and equipment they could.

On September 7, 75 and 76 Squadrons provided fighter escort to a small force of Beauforts and Beaufighters which attacked two Japanese warships and next day the final act of the Milne Bay battle was played out – an attack by Japanese bombers on the airfields which failed to achieve much material damage but did, unfortunately, kill two of 75 Squadron's ground staff.

After playing its very full part in this, the first defeat of Japanese ground forces in the Pacific war, 75 Squadron was withdrawn to Cairns in October where the exhausted Australians made the most of a short rest, prior to returning for a second stint at Milne Bay in February 1943. From its old haunt, 75 Squadron participated in its last major aerial combat of the Pacific war when, on May 14 1943, 15 Kittyhawks, from 75 and 77 Squadrons intercepted 20 Zeros, 37 Betty bombers and eight Val dive bombers heading towards Milne Bay. For the loss of one Kittyhawk, the two Australian squadrons accounted for four bombers and two fighters while another five bombers were probably destroyed.

From September a detachment of Kittyhawks began operations from Vivigani strip on Goodenough Island and in October, the rest of the unit moved to this location. A succession of further moves took 75 Squadron to a variety of bases where it flew ground attack missions against bypassed Japanese garrisons. 75 Squadron was based at Moratai by December 1944, and from here participated in its last major campaign – the invasion of Tarakan.

Returning to Australia in December 1945, 75 Squadron was based at Deniliquin where it was reduced to cadre status. A move towards the end of May 1946 bought the aircraftless fighter squadron to Schofields but it was not until a further move to Williamtown was effected on August 5 that 75 Squadron at last began to take delivery of Australian built North American Mustangs and additional personnel. With its new fighters 75 Squadron conducted air-to-air and air-to-ground combat training and participated in exercises prior to disbanding on March 25 1948.

75 Squadron reformed at Williamtown on January 24 1949 and again operated Mustangs before re-equipping with Australian built de Havilland Vampire jet fighters. Having become conversant with jet fighter operations, in July 1952 the unit, along with 76 Squadron, proceeded to Malta where it provided an air garrison to that Mediterranean island for the following two and a half

years. It is significant to note that this deployment was the first permanent overseas commitment made in peacetime by the Australian Defence Force.

While at Malta, 75 Squadron, which was flying leased RAF Vampire FB.9s and Gloster Meteors, participated in exercises centred around the defence of Malta. These were flown against and with the air forces of many countries including Great Britain, New Zealand, Canada, the United States and a number of European nations. For a period in 1953 a detachment of personnel was based in the United Kingdom where it participated in celebrations marking the Coronation of Queen Elizabeth II. Straight from this commitment, the detachment flew to Germany and participated in Exercise Coronet, up to that time the largest defence exercise held in Europe since the end of World War II.

When the detachment returned to Malta it found that the rest of 75 Squadron had moved from its earlier base at Hal Far to Takali. From here routine training continued until November 1954 when the unit disbanded, preparatory to its personnel returning to Australia.

In April 1955, 75 Squadron reformed initially with Vampires and Meteors but these were soon replaced by CAC built Sabres. With these missile armed and highly manoeuvrable jet fighters, the unit operated from Williamtown until converting to the supersonic Dassault Mirage III in August 1965. After working up to a suitable level of operational proficiency on these new aircraft, the unit moved to Malaysia in May 1967.

From Butterworth 75 Squadron participated in numerous multinational defence exercises with aircraft of the Malaysian, Singaporean, New Zealand and occasionally British air forces.

With the gradual reduction of the Australian presence at Butterworth in the mid 1980s, 75 Squadron itself was withdrawn to Australia in October 1983. Based at Darwin, it became the first fighter squadron permanently based in the Northern Territory since the end of World War II. 75 Squadron continued operating its Mirages until May 1988 when it began to re-equip with F/A-18 Hornet fighters.

Having moved to the newly reconstructed Tindal base near Katherine to the south of Darwin in October 1988, 75 Squadron today operates its multirole Hornets in the air-to-air, ground attack and anti-shipping roles. Armed with the latest generation of air-to-air and anti-shipping missiles, cannon and bombs, the rugged long range – a range which can be substantially ex-

French designed, Australian built Dassault Mirage III interceptors of 75 Squadron. (RAAF Museum)

tended with the aid of the RAAF's Boeing 707 aerial refuelling tankers – Hornets are an ideal aircraft for operations in the vast reaches of Australia's North and the unit will continue Hornet operations from Tindal for the foreseeable future.

Formed: *March 4 1942*
Squadron Code: *GA*
Aircraft: *Kittyhawk, Mustang, Vampire, Meteor, Sabre, Mirage, Hornet, Nomad*
Locations: *Townsville: March 1942*
Port Moresby: March – April 1942
Townsville: April – May 1942
Kingaroy: May – July 1942
Lowood: July 1942
Milne Bay: July – October 1942
Cairns: October 1942 – February 1943
Milne Bay: February – September 1943
Goodenough Island: Sept 1943 – Jan 1944
Nadzab: January – March 1944
Cape Gloucester: March – May 1944
Tadji: May 1944
Hollandia: May – June 1944
Biak: June – July 1944
Noemfoor: July – December 1944
Moratai: December 1944 – April 1945
Tarakan: April – December 1945
Deniliquin: December 1945 – May 1946
Schofields: May – August 1946
Williamtown: August 1946 – March 1948
Disbanded:
Williamtown: January 1949 – July 1952
Halfar: July 1952 – June 1953
Takali: June 1953 – November 1954
Disbanded:
Williamtown: April 1955 – May 1967
Butterworth: May 1967 – October 1983
Darwin: October 1983 – October 1988
Tindal: October 1988 – present

Commanded by Wing Commander P Jeffrey, 76 Squadron formed at Archerfield on March 14 1942.

The new fighter squadron, equipped with American built Curtiss P-40E Kittyhawks commenced operational training, a task completed after the unit moved to Weir strip near Townsville in mid April. While there three scrambles to intercept Japanese flying boats over Townsville were conducted but, unfortunately, the enemy were not contacted and the Kittyhawk pilots were denied an early opportunity to gain valuable combat experience.

In July 76 Squadron prepared to undertake its first operational assignment when it moved to Milne Bay. While transiting through Port Moresby, the Kittyhawk pilots took the opportunity to have an early crack at the Japanese when they carried out a strike against enemy positions at Gona. Whilst completing this mission several Japanese fighters were encountered but both sides escaped without loss during the ensuing combat.

On assembling at Milne Bay with 75 Squadron, 76 Squadron personnel were confronted by the most appalling conditions imaginable. Maintenance and domestic facilities were virtually non existent, the airfield was not finished and, with Milne Bay's extremely wet climate, aircraft were easily bogged if they ran off the primitive runways and taxiways.

Additionally, due to a lack of vehicles and handling equipment, all bogged aircraft had to be dragged out by hand. The humid, mosquito ridden climate and poor food combined to have a debilitating affect on the men's health and malaria and other tropical illness's quickly began to take a toll of personnel.

For the entire period of its World War II service, 76 Squadron operated Kittyhawks in both the fighter and fighter/bomber roles. (RAAF Museum)

Despite these difficulties, the unit settled into its new "home" and was fully established in time to meet a small Japanese raiding party of one dive bomber and four Zero fighters which arrived overhead on August 4. Eight Kittyhawks attacked and Flight Lieutenant P Ash shot down the dive bomber, Sergeants P Dempster and B Carroll shared in the destruction of a Zero and Flight Lieutenant Warn probably destroyed two more. In return a 75 Squadron Kittyhawk was destroyed on the ground.

On August 11 a dozen Zeros again struck at Milne Bay and, despite being intercepted by a much larger force of Kittyhawks from both squadrons, succeeded in shooting down several fighters from both squadrons. Flight Lieutenant Warn shot down one of the intruders, however, Flying Officer A G McLeod and Flight Sergeant G Finkster were both killed.

On the afternoon of August 25 a Japanese invasion force was detected approaching Milne Bay. The Kittyhawks, some of which had hurriedly been modified to carry bombs were "bombed up" and sent out against the Japanese fleet. Despite heavy anti-aircraft fire, the Australian fighters screamed into the attack damaging several ships. Nonetheless, with the exception of one badly damaged ship, the Japanese vessels pressed on and, under cover of darkness, proceeded into Milne Bay where they disgorged their troops.

From dawn next morning the Kittyhawks commenced intensive attacks against the enemy bridgehead, bombing and strafing barges, fuel dumps, stores and troops. The attacks were devastating in their accuracy and caused considerable loss to the invaders. In retrospect, these operations were a crucial element in the defence of Australia.

After these initial strikes, 76 Squadron flew

close support missions for the hard pressed Australian troops who were gradually being pushed back towards the airfield. Often in teaming rain, the ground staff laboured day and night rearming and refuelling aircraft which flew as many missions as possible each day.

76 Squadron was dealt a heavy blow on August 27 when its Commanding Officer, Squadron Leader Peter Turnbull, an experienced and much admired Middle East veteran, was shot down and killed by ground fire while attacking a Japanese tank. Squadron Leader Keith 'Bluey' Truscott then assumed command and it was this officer who, as the Japanese approached perilously close to the airfield and the aircraft were withdrawn to Port Moresby that night, refused to sanction a complete withdrawal of his squadron on the grounds that such a move would lower the morale of the Australian troops. Instead, he asserted, the RAAF men would join their Army comrades in the defence of the airfield with borrowed rifles.

Fortunately, such a desperate course of action was not necessary and when the Kittyhawks returned next morning they resumed their damaging attacks despite difficult weather conditions and intense anti-aircraft fire. These operations continued over the succeeding days despite the fact that the air and ground crews were by now almost totally exhausted by their exertions. For a time the Japanese advance continued and was only checked alongside one of the jungle fringed runways.

By September 1 the first signs that the Japanese were losing the will to fight on were detected and it became apparent that a nightly evacuation had commenced. The battle raged on, however, until the evening of September 7 when the last remnants of the Japanese force embarked in their ships and sailed from Milne Bay. An attack by Kittyhawks, Beauforts and Beaufighters on the retreating vessels concluded the action.

Having played a vital part in one of the most significant battles of the South West Pacific campaign – a battle which was both the first defeat of a Japanese invasion force and the most southerly penetration achieved by the Japanese during World War II – 76 Squadron was withdrawn to Darwin in October where it settled into Strauss. Despite the fact that many of its personnel were suffering from malaria and other illnesses, the unit remained operational and was responsible for the air defence of the Darwin area.

Having converted to Mustangs at the end of World War II, 76 Squadron took part in the occupation of Japan. These pilots have just arrived at Bofu after their long ferry flight. The aircraft is A68-782. (RAAF Museum)

Fortunately for the weary Australians, Japanese air activity was limited in this period. The convoy escort patrols flown were usually free of incident and the Kittyhawks were not involved in air-to-air combat by day. At night it was a different matter and on several occasions aircraft were scrambled in an attempt to intercept incoming bombers. The Kittyhawks were usually unsuccessful in these endeavours, however, Squadron Leader Truscott did manage to bring down a Betty bomber in January 1943.

In February 76 Squadron moved to Western Australia and was based at Onslow and later Potshot where it carried out routine and uneventful patrols. While at Potshot Squadron Leader Truscott, one of the RAAFs most colourful officers and the Service's second highest scoring ace, was killed on March 28 in a flying accident.

April saw 76 Squadron move to Bankstown and thence to Goodenough Island where the Kittyhawks conducted ground attack missions, barge sweeps, bomber escorts, fighter sweeps and patrols. From this time on however, the enemy was rarely encountered in the air. Successive moves took 76 Squadron to Kiriwina, Momote, Noemfoor, Moratai and Labuan where the Kittyhawks supported the invasion of Borneo.

76 Squadron's last wartime operation was flown on August 14 1945. Unlike most other fighter squadrons which were destined to be disbanded, 76 Squadron was selected for duty with the British Commonwealth Occupation Force in Japan. The first prerequisite prior to undertaking this important task was to re-equip with more modern fighter aircraft.

North American/CAC P-51D Mustangs were selected but delays in receiving the new aircraft at Labuan in turn delayed conversion training and the unit's subsequent move to Japan for several months.

Finally, in early 1946, 76 Squadron began its long awaited move on February 11 when the ground staff left by ship for Japan. The Mustangs flew out on February 26 and after a long and arduous flight, broken by stops in the Philippines and Okinawa, arrived in Japan. Based at Bofu, the unit's main role was to conduct surveillance patrols over the recently demilitarised country the Australians were now helping to occupy. Moving to Iwakuni in February 1948, 76 Squadron found its role gradually changing. With the advent of the Cold War and consequent Soviet threat to a virtually defenceless Japan, there can be little doubt that, had the Communists, made any attempt to attack Japan, that 76 Squadron would have been committed to Japan's defence. An ironic situation indeed.

As the Occupation Forces were scaled down in size, it was decided that the RAAF fighter force should be reduced from three to just one squadron. 76 Squadron was slated as one of the squadrons to go. Consequently, it disbanded at Iwakuni on October 29 1948.

76 Squadron did not remain in abeyance long and reformed at Williamtown on January 24 1949, again with Mustangs. Unfortunately, lack of personnel and aircraft severely affected operations for the first year of its existence and it was not until mid 1950 that the unit received sufficient personnel to allow it to operate effectively.

Participating in air defence exercises as well as air-to-ground and air-to-air combat training, 76 Squadron achieved a reasonable level of effectiveness prior to converting to Vampire F30 jet fighters during 1952.

In July, 76 Squadron sailed for Malta where it, along with 75 Squadron, became responsible for Malta's air defence. Initially based at Halfar the unit flew leased British Vampire FB.9s which differed somewhat in performance to the Australian manufactured Vampires previously operated.

For the next two years 76 Squadron remained at Malta and later operated from Takali airfield. The unit participated in a considerable number of air defence exercises with British, American, Canadian and European air forces. By Christmas 1954, 76 Squadron's tour of duty was all but over and the unit handed over its aircraft and equipment to the British prior to sailing for Australia in January 1955. Returning to Williamtown, 76 Squadron disbanded on March 16 1955.

76 Squadron reformed at Williamtown on January 11 1960. Initially equipped with Vampires, a year and a half later these aircraft were exchanged for Sabres. The highly manoeuvrable Sabres proved to be so agile that later, 76 Squadron formed its own aerobatic teams – the Red Diamonds and the Black Panthers. These teams participated in numerous airshows and public events, the "Black Panthers" aircraft in particular being especially striking with the large leaping panther painted under the cockpit of each team member's jet.

76 Squadron was launched into the supersonic age during August 1966 when it began Dassault Mirage III operations. With these French designed/Australian built aircraft, 76 Squadron continued in its air defence and ground attack roles for several years until a review of RAAF fighter operations and post Vietnam economy measures forced the unit's disbandment on August 24 1973.

Having again reformed at Williamtown on January 1 1989, 76 Squadron today operates Macchi MB-326 jets and (up until late 1995) Winjeel FAC aircraft. The Italian designed/Australian built Macchis provide jet experience for pilots selected for "fast jet" (Hornet/F-111) flying later in their careers. Other roles for the Macchis include army and navy support tasks. The Winjeels on the other hand were utilised in the forward air control role, marking targets with smoke grenades to assist close support aircraft in making their attacks.

While 76 Squadron will continue its operations from Williamtown, its aircraft are set to change. From mid 1995 specially modified Pilatus PC-9s replaced the Winjeels in the forward air control role. Additionally, the RAAF is looking for a suitable jet trainer/lead-in fighter to replace the aging Macchi. Any replacement must have a hard hitting ground attack and close support capability – a capability long sought by the Army.

A 76 Squadron Macchi at Fairbairn.
(Glenn Alderton, RAAF)

Once re-equipped, 76 Squadron will provide an important operational capability additional to its training role – a role which will continue for the foreseeable future.

Formed: *March 14 1942*
Squadron Codes: *I, SV*
Aircraft: *Kittyhawk, Mustang, Vampire, Sabre, Mirage, Macchi, Winjeel, PC-9*
Locations: *Archerfield: March – April 1942*
Townsville: April – July 1942
Milne Bay: July – October 1942
Strauss: October 1942 – February 1943
Onslow: February – March 1943
Potshot: March – April 1943
Bankstown: April – May 1943

Goodenough Island: May – August 1943
Kiriwina August: 1943 – March 1944
Momote: March – September 1944
Noemfoor: September 1944 – April 1945
Moratai: April – June 1945
Labuan: June 1945 – March 1946
Bofu: February 1946 – March 1948
Iwakuni: March – October 1948
Disbanded:
Williamtown: January 1949 – July 1952
Halfar: July 1952 – June 1953
Takali: June 1953 – January 1955
Williamtown: January – March 1955
Disbanded:
Williamtown: January 1960 – August 1973
Disbanded:
Williamtown: January 1989 – present

------

77 SQUADRON

77 Squadron formed at Pearce on March 16 1942. Equipped with Curtiss P-40E Kittyhawk fighters, the unit, initially under the command of Squadron Leader R Brooker, moved to Guildford in late April where it was made responsible for the air defence of Perth. At that time the Western Australia capital was expected to come under attack from Japanese carrier borne aircraft.

In August 77 Squadron moved to Batchelor near Darwin where it became the first RAAF fighter squadron stationed in the area. After a further move to Livingstone, the Kittyhawks made several attempts to intercept Japanese bombers raiding Darwin after dark. On the night of 23/24 November 77 Squadron obtained its first kill when Squadron Leader R C Cresswell attacked three Betty bombers silhouetted against the moon and shot one down in flames.

February 1943 saw 77 Squadron join 75 Squadron at Milne Bay, which at that time was receiving little attention from the enemy. There were exceptions to this quiet existence and on April 11 two Kittyhawks attacked several Japanese fighters destroying one and damaging another. Later, on May 14, a force of 65 Japanese aircraft raided Milne Bay and were engaged by 15 Kittyhawks from both Australian squadrons. In the ensuing combat four bombers and two fighters were shot down and a further five bombers probably destroyed. One Kittyhawk was lost.

After moving to Vivigani strip on Goodenough Island 77 Squadron flew fighter escort missions to bombers attacking Gasmata and commenced ground attack operations using the Kittyhawk's

fifty calibre machine guns and bombs. Barge sweeps were also a regular feature of the unit's operations.

77 Squadron relocated to Los Negros in early 1944, the RAAF ground staff going ashore just after the beachhead was seized from the Japanese. After flying some ground attack missions in support of the invasion forces, the unit moved to Noemfoor in September and subsequently to Moratai in April 1945. From here, ground attack missions against bypassed Japanese garrisons continued until June when 77 Squadron moved to Labuan to support the invasion of Borneo.

After Japan's surrender 77 Squadron was selected to form part of the Allied occupation force which was to garrison that country. In September 1945 it began to convert to Mustangs and proceeded to Japan in February/March 1946. Initially based at Bofu and later Iwakuni, 77 Squadron flew surveillance patrols over large areas of Japan and engaged in normal peacetime flying training.

Over four years later when the Korean War broke out on June 25 1950, 77 Squadron was preparing to leave Japan on the completion of its occupation duties. These plans were dashed within days when the Australian Mustangs were committed to support United Nation forces in Korea. Through the commitment of 77 Squadron, Australia became the first United Nations country outside the United States to conduct combat operations in the defence of South Korea.

Operations commenced on July 2 1950 when uneventful escort missions were flown in support

Armed with rockets, Meteors of 77 Squadron prepare to takeoff on a Korean War ground attack mission. With its much maligned Meteors 77 Squadron scored the RAAF's last air-to-air victories during the Korean War. (RAAF Museum)

of American bombers and transport aircraft. Thereafter, the Mustangs were used extensively in the close support and interdiction roles, striking Communist targets in South and, later, North Korea. On July 7 Australia suffered its first casualty of the Korean War when Squadron Leader G Strout was shot down by anti-aircraft. He was the first of 42 77 Squadron pilots to die on active service in this campaign.

In the early months of the Korean War, the Australian airmen played a role totally out of proportion to the number of aircraft they possessed. American squadrons were urgently equipped with Lockheed F-80 Shooting Star jet fighters. The F-80s had some difficulty in performing ground attack work – the main task required of the United Nations Air Forces at this stage of the war – due to performance limitations inherent in their design. Consequently, 77 Squadron found itself as the only fighter squadron equipped with aircraft capable of providing effective support to the ground forces for some time until the United States could dispatch its own Mustangs, Corsairs and Skyraiders to Korea.

After some ten months of intensive ground attack operations, in which the Mustangs had

operated from various Korean airfields, 77 Squadron was withdrawn to Iwakuni in April 1951 to re-equip with Gloster Meteor twin engined jet fighters. Towards the end of July the unit was operational once more and moved to Kimpo on the Korean mainland where, on July 30, it flew its first Meteor combat mission – an uneventful fighter sweep. Operations of this type, penetrating deeply into North Korean airspace, continued for several weeks without meeting the enemy, a situation that was soon to change.

77 Squadron had its first major engagement against enemy aircraft when, on August 29, eight Meteors were attacked by 30 Chinese MiG-15 jets. One Meteor, piloted by Warrant Officer Guthrie, was shot down from an altitude of 34,700 feet and the pilot forced to eject (creating an unofficial world height record for ejection in the process) and spent the rest of the war as a POW. The remaining Meteors – including one badly damaged jet – escaped but did not inflict any damage on the enemy aircraft.

The Australians gained their first confirmed MiG kill on December 1 when 12 Meteors were engaged by over 50 MiG-15s over Pyongyang. Flying Officer B Gogerly shot down one MiG and the unit as a whole claimed a second kill but three Meteors were also lost – two of their pilots ejecting to survive as POWs – while a further two aircraft were damaged.

After this engagement, the Meteor's role was switched to ground attack work – a role better

Sidewinder armed Sabres of 77 Squadron at readiness wait at their Malaysian airfield during the "Confrontation" with Indonesia.

suited to the Meteor's capabilities. In this capacity, the Meteors took a considerable toll on North Korean and Chinese ground forces but themselves suffered heavy losses from anti-aircraft fire, MiG attacks and the dangerous flying conditions prevalent in Korea due to the savage nature of the climate and rugged terrain.

Ground attack operations continued until the end of the war but from mid 1952, the Meteors occasionally met MiG-15s at low level. In these engagements Pilot Officer J Surman probably shot down a MiG on May 4 1952 while four days later Pilot Officer W Simmonds definitely destroyed another. 77 Squadron's last brush with enemy aircraft occurred on March 27 1953 when several MiGs and four Meteors clashed. One Meteor was damaged but Sergeant G Hale probably destroyed one enemy jet and damaged another.

The Korean War ended on July 27 1953 although 77 Squadron remained on garrison duty for some time to come. In March 1954, the unit moved to Kunsan and in October, back to Iwakuni prior to returning to Australia in November. 77 Squadron, which had been away from Australia for 11 years, disbanded at Williamtown on August 12 1956.

Equipped with Australian built CA-27 Sabres, 77 Squadron reformed at Williamtown on November 19 1956. In December 1958, the unit moved to Butterworth, Malaya. British Commonwealth forces were still engaged in anti-terrorist operations at the time and the Sabres flew a few ground attack missions against jungle covered targets before the "Malayan Emergency" was officially concluded in mid 1960.

77 Squadron remained at Butterworth throughout most of the 1960s providing an air defence capability for Malaysia during the period of Confrontation with Indonesia, prior to returning to Williamtown in early 1969 to convert to the Dassault Mirage IIIs. These highly regarded aircraft remained in service for almost 20 years and participated in numerous air defence exercises both within Australia and overseas.

In January 1985, 77 Squadron took over the MB-326 Macchi jet trainers of 2 Operational Conversion unit while that unit converted to McDonnell Douglas F/A-18 Hornets. 77 Squadron itself re-equipped with these highly capable multirole fighters in 1987.

Today 77 Squadron operates in the air defence, ground attack and anti-shipping roles and conducts regular deployments to Malaysia, New Zealand, the Philippines and other friendly Pacific nations. The Hornets equipping 77 Squadron are planned to remain in service for at least another 20 years and so equipped, 77 Squadron will remain at the forefront of Australia's air defence.

Formed: *March 16 1942*
Squadron Code: *AM*
Aircraft: *Kittyhawk, Mustang, Meteor, Sabre, Mirage, Macchi, Hornet*
Locations: *Pearce: March – April 1942*
Guildford: April – August 1942
Batchelor: August – September 1942
Livingstone: September 1942 – February 1943
Milne Bay: February – June 1943
Goodenough Island: June 1943 – February 1944
Los Negros: February – September 1944
Noemfoor: September 1944 – April 1945
Moratai: April – June 1945
Labuan: June 1945 – March 1946
Bofu: March 1946 – March 1948
Iwakuni: March 1948 – October 1950
Pohang: October – November 1950
Yonpo: November – December 1950
Pusan: December 1950 – April 1951
Iwakuni: April – July 1951
Kimpo: July 1951 – March 1954
Kunsan: March – October 1954
Iwakuni: October – November 1954
Williamtown: November 1954 – August 1956
Disbanded:
Williamtown: November 1956 – December 1958
Butterworth: December 1958 – February 1969
Williamtown: February 1969 – present

---

78 SQUADRON

78 Squadron, equipped with Kittyhawks, formed at Camden on July 20 1943 under the command of Squadron Leader G F Walker.

Operational by October, the unit began moving to Woodlark Island, however this was changed to Kiriwina in November. From here bomber escort – both to RAAF and USAAF aircraft – and ground attack missions, often against targets around Gasmata, were flown.

78 Squadron moved to Nadzab in January 1944, where its Kittyhawks supported Australian Army operations by flying close support missions as well as attacking Japanese stores areas and occupied villages.

A Mustang of 78 Squadron crosses the rugged Japanese coastline.

78 Squadron moved to Cape Gloucester in March where ground attack operations continued. In April the unit relocated to Tadji and this was followed by a further move to Hollandia in May. While at Hollandia, 78 Squadron fought the RAAF's last major air combat of the Pacific war when, on June 3 1944, 16 Kittyhawks attacked a formation of 12 Oscar fighters and three Kate dive bombers. One Australian pilot, Flight Sergeant W H Harnden, was killed but in return the Kittyhawks claimed the destruction of nine enemy aircraft. Most of the surviving Japanese aircraft were damaged. The most successful pilot was Flying Officer H G White who destroyed an Oscar, a Kate and damaged a second Kate.

On June 10 Flight Lieutenant D R Baker destroyed a Tony fighter. This was another significant event as it represented the RAAF's last aerial victory of the New Guinea campaign. By the end of the month, 78 Squadron had moved again, this time to Noemfoor Island, and continued ground attack missions and barge sweeps while also supporting US Navy PT boat operations. 78 Squadron relocated to Moratai in December 1944 and from here conducted strikes against targets in the Halmaheras and Celebes.

With the seizure of Tarakan, 78 Squadron began operations from Tawi Tawi strip from July 20 and was here at war's end. After the cessation of hostilities 78 Squadron flew some leaflet dropping and reconnaissance flights over Japanese territory prior to returning to Australia in December.

Based initially at Deniliquin and later Schofields, 78 Squadron was reduced to cadre status until August 1946 when it moved to Williamtown and became operational with Mustangs. This second phase of the unit's life was a short one, however, and after participating in a number of naval and army co-operation exercises, 78 Squadron officially disbanded on April 1 1948.

Formed: *July 20 1943*
Squadron Code: *HU*
Aircraft: *Kittyhawk, Mustang*
Locations: *Camden: July – November 1943*
Kiriwina: November 1943 – January 1944
Nadzab: January – March 1944
Cape Gloucester: March – April 1944
Tadji: April – May 1944
Hollandia: May – June 1944
Noemfoor: June – December 1944
Moratai: December 1944 – July 1945
Tarakan: July – December 1945
Deniliquin: December 1945 – May 1946
Schofields: May – August 1946
Williamtown: August 1946 – April 1948
Disbanded: *April 1 1948*

---

79 SQUADRON

Equipped with Supermarine Spitfire VCs, 79 Squadron formed at Laverton on April 26 1943 under the command of Squadron Leader A C Rawlinson. Almost immediately, however, the unit moved to Wooloomanata, near Geelong.

By June, 79 Squadron was operational and deployed to Goodenough Island. From here, the first scramble occurred on June 19 1943, however, no contact with the enemy was made.

In August, the unit moved to Kiriwina in the Trobriand Islands. Conditions were primitive and maintenance facilities crude. Despite these difficulties, October 31 saw 79 Squadron obtain its first kill when Flight Sergeant I H Callister shot down a Tony north of Kiriwina.

In late November 79 Squadron commenced fighter sweeps over enemy occupied areas and on the 28th, Flying Officer A W Moore, who was carrying out a test flight, located and shot down a Japanese Dinah reconnaissance aircraft. A second Dinah was shot down by a Spitfire carrying out another test flight on November 28.

Early 1944 saw the unit undertake more active operations when ground attack sorties commenced. Defensive patrols, fighter sweeps, bomber escorts and naval co-operation tasks were also flown. In March 79 Squadron relocated to Momote Island which had just been captured from the Japanese. Despite difficult conditions and a lack of spare parts for the Spitfires, ground attack missions continued to be flown.

79 Squadron moved south to Darwin in January 1945, but next month was again in action, this time from Moratai. After Japan's surrender, leaflet dropping operations were conducted over enemy occupied islands prior to the unit moving to Oakey where it disbanded on November 12 1945.

79 Squadron reformed from elements of 77 Squadron at Singapore (with Sabres) on May 28 1962 and proceeded that same day to Ubon, Thailand, where it was destined to be based for six years. The unit's role was to help resist an anticipated invasion of Thailand by North Vietnamese forces. In a unique arrangement, aircrew and technical personnel rotated every three months with the fighter squadrons at Butterworth while non technical personnel were sent from Australia for normal posting periods.

On arrival at Ubon, 79 Squadron was immediately placed at a state of operational readiness and patrolling aircraft, as well as those on ground alert, were kept fully armed. No attack eventuated, however, and the closest the Sabres came to seeing combat was their participation in several SEATO exercises. However, the presence of 79 Squadron contributed significantly to Thailand's security. By the late 1960s, however, that country's defence seemed assured and the presence of the Sabres was no longer necessary. Consequently, 79 Squadron disbanded at Ubon on July 31 1968.

On March 31 1986, 79 Squadron reformed at Butterworth. The unit's 10 Mirage jet fighters and personnel were inherited from 3 Squadron which was returning to Australia to re-equip with Hornets. 79 Squadron participated in numerous air defence exercises and was the RAAF's last permanent fighter presence in Malaysia. 79 Squadron had only been formed as a temporary measure during a transitional period when the Service's other fighter squadrons progressively

Appearing relatively undamaged after crash landing is this 79 Squadron Spitfire. (RAAF Museum)

became nonoperational to re-equip with Hornets.

Once this had been completed 79 Squadrons role was fulfilled and the unit disbanded on June 30 1988.

Formed: *April 26 1943*
Squadron Code: *UP*
Aircraft: *Spitfire, Sabre, Mirage*
Locations: *Laverton: April – May 1943*
Wooloomanata: May – June 1943
Goodenough Island: June – August 1943
Kiriwina: August 1943 – March 1944
Momote: March 1944 – January 1945
Darwin: January – February 1945
Moratai: February – October 1945
Oakey: October – November 1945
Disbanded:
Singapore: May – May 1962
Ubon: June 1962 – July 1968
Disbanded:
Butterworth: March 1986 – June 1988
Disbanded: *June 30 1988*

A 79 Squadron Mirage launches a Matra missile in Malaysian skies.

80 SQUADRON

Equipped with Kittyhawks, 80 Squadron formed at Townsville on September 10 1943.

The new fighter squadron, under the command of Squadron Leader G A Cooper, relocated to Aitkenvale the following month where flying training commenced. Operational by February 1944, the unit moved to Nadzab in New Guinea where combat operations commenced on February 26.

Engaged mainly in bomber escort and ground attack operations, 80 Squadron operated from a succession of bases including Cape Gloucester (March), Tadji (April), Hollandia (May), Biak and Noemfoor (both in July). Prior to moving to Biak, many attack missions against that island were flown and also against Wakde where the Kittyhawks attacked beachhead targets while American assault troops landed.

By the end of 1944 worthwhile targets in 80 Squadron's area of operations were few and, along with other squadrons in a similar situation, morale began to suffer. By July 1945, 80 Squadron was operating from Tarakan but targets warranting air attack were still difficult to find.

After the cessation of hostilities, 80 Squadron flew reconnaissance missions over Japanese occupied territory prior to returning to Australia in December. 80 Squadron disbanded at Deniliquin on July 11 1946.

Formed: *September 10 1943*
Squadron Code: *BU*
Aircraft: *Kittyhawk*
Locations: *Townsville: Sept – October 1943*
Aitkenvale: October 1943 – February 1944
Nadzab: February – March 1944
Cape Gloucester: March – April 1944
Tadji: April – May 1944
Hollandia: May – July 1944
Biak: July 1944
Noemfoor: July 1944 – January 1945
Moratai: January – July 1945
Tarakan: July – December 1945
Deniliquin: December 1945 – July 1946
Disbanded: *July 11 1946*

A formation of 80 Squadron Kittyhawks. (RAAF Museum)

82 SQUADRON

82 Squadron formed at Bankstown on June 18 1943 under the command of Squadron Leader S W Galton. The unit was to be equipped with Kittyhawk fighters but, due to an initial shortage of these aircraft, one flight temporally received Bell P-39 Airacobras.

82 Squadron commenced a move to the Darwin area in April 1944, but this was changed to Townsville's Ross River Strip. From here training continued with emphasis on dive bombing, navigation, gunnery and naval co-operation.

Having moved to Noemfoor in August 1944, on September 7 two Kittyhawks flew 82 Squadron's first combat mission when they bombed and strafed targets at Sorong. Operating initially from Kamiri strip, 82 Squadron undertook ground attack missions as well as barge sweeps along the coast. Operational effort at this time proved difficult to sustain as most of the ground staff and much equipment remained at Townsville until November awaiting forward movement.

In early 1945, 82 Squadron moved to Moratai where it continued its ground attack missions against bypassed Japanese garrisons on nearby islands. Additionally, considerable operational effort was expended on convoy patrols around Borneo.

June saw 82 Squadron move once again, this

82 Squadron Mustangs at Labuan shortly before the unit moved to Japan. (RAAF Museum)

time to the recently captured Labuan Island, where it provided close air support to Australian troops during the Borneo operations.

82 Squadron's most successful strike in this period occurred in mid August when several Kittyhawks bounced Japanese aircraft about to take-off from Kuching airfield. Four were destroyed and two others damaged before the fighters turned their attention to barge traffic on the Sarawak River.

Almost as soon as hostilities ceased, 82 Squadron was informed that it would form part of the British Commonwealth Occupation Force which was to garrison a defeated Japan. The unit converted to Mustangs and, in February/March 1946, moved to Bofu in Japan. The move was not without incident as three Mustangs and an escorting

Mosquito crashed into the sea just 100km short of their new base after running into extremely bad weather. All the aircrews involved in this incident perished.

As part of the Occupation Force, 82 Squadron was engaged on surveillance patrols over Japan as well as exercises, firepower demonstrations and flypasts. These tasks continued for the next two years, however, by 1948 Australia had decided to reduce its forces in Japan and, accordingly, 82 Squadron disbanded at Iwakuni on October 22 1948.

Formed: June 18 1943
Squadron Code: FA
Aircraft: Kittyhawk, Airacobra, Mustang
Locations: Bankstown: June 1943 – April 1944
Ross River: April – August 1944
Noemfoor: August 1944 – April 1945
Moratai: April – June 1945
Labuan: June 1945 – March 1946
Bofu: March 1946 – March 1948
Iwakuni: March – October 1948
Disbanded: October 22 1948

83 SQUADRON

Under the command of Squadron Leader W J Meehan, 83 Squadron formed at Strathpine on February 26 1943. Initially equipped with six Bell P-39D Airacobra fighters, the unit later received Boomerangs as replacements.

From Strathpine, routine patrols were conducted as well as the occasional interception of unidentified aircraft – which invariably proved to be friendly – before moving to Melville Island in December 1943. In January 1944 a further move, to Gove was effected where the Boomerangs provided fighter cover for Allied shipping and conducted patrols.

Both tasks were uneventful.

The enemy had not been met before 83 Squadron moved to Camden in August 1944. A further move to Menangle was made where 83 Squadron disbanded on September 18 1945.

Formed: February 26 1943
Squadron Code: MH
Aircraft: Airacobra, Boomerang
Locations: Strathpine: Feb 1943 – Dec 1943
Melville Island: December 1943 – January 1944
Gove: January – August 1944
Camden: August 1944 – February 1945
Menangle: February – September 1945
Disbanded: September 18 1945

(right) Landing mishap for 83 Squadron's Boomerang "MH-T". (RAAF Museum)

84 SQUADRON

On February 5 1943, 84 Squadron, commanded by Squadron Leader N Ford, formed at Richmond.

Equipped with Boomerangs the unit moved to Horn Island on the tip of Cape York in April, where, along with the Kittyhawks of 86 Squadron, it became responsible for the air defence of Horn Island and Merauke. Due to a lack of enemy activity, however, this was not an onerous task and flying mainly consisted of uneventful patrols over Merauke – where a flight was later based.

On May 16 1943 two Boomerangs flown by Flying Officer R W Johnstone and Sergeant M F J Stammer whilst on a routine patrol located and attacked three Betty bombers. The Australian built fighters and Japanese bombers exchanged fire, however, the Bettys escaped into cloud before hits could be registered on them.

There were one or two notable exceptions to this peaceful existence. On September 9 1943, four Boomerangs and a larger force of 86 Squadron Kittyhawks scrambled to intercept a Japanese bomber raid against Merauke. While the Kittyhawks shot down some of the enemy aircraft, the slower Boomerangs were unable to close for combat. One Boomerang was destroyed on the ground.

84 Squadron was allocated Kittyhawks in September 1943 and, with these aircraft, flew a few ground attack missions against targets in Dutch New Guinea. The unit moved to Aitkenvale in May 1944 where it was reduced to cadre status

pending re-equipment with Mustangs. Next month a move to Macrossan was effected where, in May 1945, 84 Squadron received its first Mustangs.

Before the unit had completed its conversion training, the war ended and 84 Squadron disbanded at Townsville on January 29 1946.

Formed: *February 5 1943*
Squadron Code: *LB*
Aircraft: *Boomerang, Kittyhawk, Mustang*
Locations: *Richmond: February – April 1943*
Horn Island: April 1943 – May 1944
Aitkenvale: May – June 1944
Macrossan: June – November 1944
Ross River: November 1944 – January 1946
Disbanded: *January 29 1946*

A formation of 84 Squadron Kittyhawks.

85 SQUADRON

85 Squadron formed at Guildford on February 12 1943 under the command of Squadron Leader C N Daly. The new fighter squadron, for want of a better aircraft, was initially allocated Brewster Buffalo Mk.Is, a fighter which had earlier given a very poor showing against the Japanese in Malaya. Improved aircraft were on order and at the end of April Australian built Boomerangs arrived. Almost immediately, a permanent detachment of six aircraft began operations from Potshot.

With its Boomerangs, 85 Squadron commenced an intensive flying training program, concentrating on formation flying, 'shadow shooting', aerobatics and fighter tactics. An unusual facet of the unit's operations was the refuelling and maintenance of civilian aircraft at Guildford. This was made an 85 Squadron task as the Federal Government considered that Perth did not, at that stage, warrant a civil airport.

Due to the almost nonexistent level of Japanese activity on Australia's West Coast, 85 Squad-

(right) An uncamouflaged Spitfire VC of 85 Squadron at Pearce in 1944. (RAAF Museum)

ron was destined never to see action and there were few alerts to test the unit.

The Potshot detachment came closer to engaging the enemy when, on May 20 1943, two Boomerangs were scrambled in an unsuccessful attempt to intercept approaching unidentified aircraft. The next day, unidentified aircraft dropped nine bombs in the sea near the detachment's camp. Two Boomerangs took off to intercept and one of them almost caught the raiders but then had engine trouble which enabled them to escape. Again, in July, an alert was sounded at Potshot but there was no contact with the enemy.

In October 1943 unit routine was disrupted by an emergency deployment to Derby made in anticipation of a Japanese naval incursion into the Indian Ocean which did not eventuate. September 1944 saw 85 Squadron begin its conversion to Spitfire Vs.

On May 16 1945 the unit moved to Pearce and in May provided fighter escort for the aircraft carrying the remains of Australia's late Prime Minister, Mr John Curtin, which were being returned to Western Australia for burial. 85 Squadron disbanded at Pearce on December 10 1945.

Formed: *February 12 1943*
Squadron Code: *SH*
Aircraft: *Buffalo, Boomerang, Spitfire*
Locations: *Guildford: Feb 1943 – May 1945*
Pearce: May – December 1945
Disbanded: *December 10 1945*

86 SQUADRON

86 Squadron, under the command of Squadron Leader W J Meehan, formed at Gawler on March 4 1943.

Allocated Kittyhawk fighters, the unit moved to the Townsville area in May where it completed its operational training. In the first few days of July 1943 the new Kittyhawk squadron deployed to Marauke, Dutch New Guinea, where it was to guard against enemy air activity in the area.

Unfortunately for 86 Squadron's personnel, there was little opportunity to meet the enemy. A chance to engage in something more than uneventful patrols occurred on July 27 when four enemy aircraft were detected by radar approaching Marauke. Sixteen Kittyhawks scrambled to intercept but failed to close with the enemy due to the late warning given. Disappointed pilots returned to base but a week later, on September 9, the unit achieved some success when 14 Kittyhawks engaged 16 Betty bombers and 16 fighters as they raided the Australian airfield. An 84 Squadron Boomerang was destroyed on the ground and most of the airborne Kittyhawks suffered from gun malfunctions. Despite this disability though, Flight Lieutenant C W Stark and Flying Officer A D Tucker each shot down a Zero while Flying Officer H W Stuart bought down an Oscar.

This combat was 86 Squadron's only major aerial battle of the war. Some ground attack missions were flown but these were generally unspectacular in nature. One relatively successful strike, however, occurred on January 31 1944 when four Japanese barges in the mouth of the Lorentz River were caught and sunk by the Kittyhawk pilots.

Further, albeit limited, air combat opportunities came 86 Squadron's way in the first weeks of 1944 when plans were laid to intercept patrolling Japanese aircraft known to make landfall over Cape Valsch. On January 22 Flight Lieutenant Stark shot down a Betty bomber and the following day Flight Lieutenant R J C Whittle and Flight Sergeant S T Kerrison caught a Betty and a Zero in the same area and destroyed both aircraft. Thereafter, the Japanese abandoned their reconnaissance flights in this area.

86 Squadron's future was somewhat in jeopardy

Kittyhawks of 86 Squadron. (RAAF Museum)

a few months later when its Kittyhawks were allocated to units engaged in more intensive operations against the Japanese. Reduced to cadre status and transferred to Bohle River, the unit later received another allocation of Kittyhawks and continued routine training until June 1945 when it began a long delayed re-equipment with Mustangs.

Plans to move forward into a combat area were, however, forestalled by the Japanese surrender. For a short period 86 Squadron continued its training and maintained detachments at Thursday Island and Marauke, before disbanding on December 20 1945.

Formed: *March 4 1943*
Squadron Code: *MP*
Aircraft: *Kittyhawk, Mustang*
Locations: *Gawler: March – May 1943*
Ross River: May – July 1943
Marauke: July 1943 – May 1944
Bohle River: May – June 1944
Macrossan: June 1944 – January 1945
Bohle River: January – December 1945
Disbanded: *December 20 1945*

87 SQUADRON

Under the command of Squadron Leader L Law, 87 Squadron formed at Coomalie Creek on September 10 1944.

The new unit, which had previously been known as 1 Photographic Reconnaissance Unit, was initially equipped with one de Havilland Mosquito and two CAC Wirraways and was tasked to provide photo reconnaissance support for Allied squadrons operating against the Japanese from bases in northern Australia.

Despite the limited number of aircraft available, reconnaissance missions were flown over Koepang, Macassar and the Masesl Islands and, as more Mosquitoes were received, the unit's rate of effort increased. 87 Squadron's activities were severely curtailed in December due to the onset of the wet season which affected photographic work for the next three months.

Apart from the usual photo reconnaissance flights, April 1945 saw 87 Squadron tasked to shadow the Japanese cruiser *Isuzu* and four smaller warships. During one of these flights a Japanese fighter attempted to intercept a Mosquito however, when the Australian crew observed the enemy approaching, they simply increased speed and left it behind. This was the

only recorded instance of enemy aircraft attempting to intercept an 87 Squadron Mosquito. Following ineffectual attacks by RAAF and Dutch aircraft, *Isuzu* was later sunk by an American submarine.

In June a detachment of three Mosquitoes deployed to Cocos Island to photograph targets in Singapore. Unfortunately, bad weather and the loss of one aircraft resulted in the detachment being a failure. Only one sortie – to Christmas Island and not to Singapore – was successfully completed.

On June 2 1944 a significant mission was flown by Flight Lieutenant N G Johnston and Flying Officer O H L Williamson. The crew located a number of Japanese bombers hidden in a gully near Cape Chater airfield on Timor. As a result of this information, an attack, which resulted in the destruction of the aircraft, was made by Spitfires operating from Darwin.

87 Squadron's longest reconnaissance of the war was carried out on July 23 when a Mosquito photographed Semarang in Java, a distance of 3700 kilometres. Low level reconnaissance missions were also flown for the first time during July.

August 15 1945 saw 87 Squadron fly its last wartime mission and after the Japanese surrender reconnaissance missions were flown over enemy occupied islands to monitor Japanese dispositions until they could be disarmed. In October 87 Squadron moved to Parkes where it disbanded on July 24 1946.

After disbanding, the Mosquitoes were allocated to the Survey Flight at Fairbairn. In November 1946 the Survey Flight was given squadron status and on March 8 1948 retitled 87 Squadron.

Operating in the photo survey role, 87 Squadron carried out many important operations in conjunction with the Commonwealth Survey Committee and National Mapping Council.

In the latter part of 1949, photo reconnaissance tasks again became a unit responsibility.

In January 1951 a Mosquito was despatched to photograph the devastation caused by the Mount Lamington Volcano eruption in New Guinea. The Mosquito crew bought back photographs of this disaster which claimed the lives of upwards of 4000 people.

March 1953 saw the commencement of 87 Squadron's final task with a survey of the Great Sandy Desert. The withdrawal of funding by the Department of National Development made the future of 87 Squadron uncertain and the unit finally disbanded at Fairbairn in December 1953.

87 Squadron Mosquitos and personnel at Port Hedland in 1953. (RAAF Museum)

Formed: September 10 1944
Squadron Code: QK
Aircraft: Wirraway, Mosquito
Locations: Coomalie Creek: September 1944 – October 1945
Parkes: October 1945 – July 1946
Disbanded:
Fairbairn: March 1948 – December 1953
Disbanded: December 1953

92 SQUADRON

Commanded by Squadron Leader H D Foot, 92 Squadron formed at Kingaroy on May 25 1945.

The first Beaufighters for the new attack squadron arrived early in July and training commenced, however, hostilities ceased before the unit could become operational. The unit disbanded on September 17 1945.

Formed: May 25 1945
Squadron Code: OB
Aircraft: Beaufighter
Locations: Kingaroy: May – September 1945
Disbanded: September 17 1945

93 SQUADRON

Commanded by Squadron Leader D K H Gulliver, 93 Squadron formed at Kingaroy on January 22 1945. After three months training with its Beaufighters, the unit prepared to depart for Labuan to support the invasion of Borneo. Unfortunately, serious delays in preparing Labuan's recently captured airfield delayed the unit's commitment to the campaign.

93 Squadron's first mission was flown on July 26 when two aircraft flew an armed reconnaissance over Borneo. Five days later, rocket armed Beaufighters sank a Japanese oil tanker with an amazing 19 direct hits and nine near misses – clearly the unit's operational training had been very effective. On the way home, a large barge was sunk killing a number of Japanese troops.

A 93 Squadron Beaufighter. The unit was nicknamed the "Green Ghosts".

One Beaufighter was lost during this operation, however, the crew was later recovered.

Dubbed the Green Ghost Squadron, a few more 87 Squadron missions were flown before the Japanese surrender, after which the Beaufighters were used to drop leaflets to the Japanese and Allied POWs, advising them of Japan's capitulation. A later task was the provision of Beaufighters to provide navigation escorts to formations of RAAF single engined fighters returning to Australia.

In early 1946 several aircraft provided navigation escorts to 81 Wing Mustangs deploying to Japan as part of the British Commonwealth Occupation Force. Returning to Australia in January 1946, 93 Squadron disbanded at Narromine on August 22 1946.

Formed: *January 22 1945*
Squadron Code: *SK*
Aircraft: *Beaufighter*
Locations: *Kingaroy: January – July 1945*
Labuan: July 1945 – January 1946
Narromine: January – August 1946
Disbanded: *August 22 1946*

94 SQUADRON

94 Squadron formed at Castlereagh on May 30 1945 under the command of Squadron Leader V H Hunt.

Allocated an attack role, training was severely hampered due to the slow delivery of the unit's Australian built de Havilland Mosquito FB.40 aircraft. Operational training was still in progress when the war ended and 94 Squadron disbanded at Richmond on January 27 1946.

Formed: *May 30 1945*
Squadron Code: *MX*
Aircraft: *Mosquito*
Locations: *Castlereagh: May 1945 – Jan 1946*
Richmond: January 1946
Disbanded: *January 27 1946*

99 SQUADRON

99 Squadron formed at Leyburn on February 1 1945 under the command of Squadron Leader J H Marshall. The first Consolidated B-24 Liberator heavy bombers arrived at Leyburn on March 8 and later in the month the unit moved to Jondaryan where training continued.

Declared operational by April, 99 Squadron's 700 personnel and 14 Liberators began the move to Darwin the following month but this had not been fully completed prior to Japan's surrender. Despite the unexpected end of the war, 99 Squadron did carry out some useful work as, with the arrival of Australian POWs at Darwin, the Liberators ferried some of these men south to major capital cities to be reunited with their families. A total of 495 POWs and 299 other personnel were moved south in this manner and on the return flights 38,757 kilograms of freight was delivered to the Northern Territory.

Having completed its only war related activity, 99 Squadron moved to Tocumwal in November 1945 and disbanded on June 5 1946.

Formed: *February 1 1945*
Squadron Code: *UX*
Aircraft: *Liberator*
Locations: *Leyburn: February – March 1945*
Jondaryan: March – May 1945
Darwin: May – September 1945
Tocumwal: September 1945 – June 1946
Disbanded: *June 5 1946*

100 SQUADRON

Named in honour of an RAF squadron destroyed flying hopelessly obsolete Vickers Vilderbeest biplane torpedo bombers in the Malayan campaign, the RAAF's 100 Squadron formed on February 15 1942 out of a personnel nucleus of the original 100 Squadron.

Initially based at Richmond, and commanded by Wing Commander A W D Miller, the unit equipped with Bristol Beauforts and moved to Mareeba in May where it conducted further torpedo bomber training and anti-submarine patrols. After reaching operational proficiency and having further moves to Laverton and Bohle River, 100 Squadron deployed to Milne Bay in October.

Prior to this, in June a detachment of Beauforts temporarily based at Port Moresby conducted the RAAF's first Beaufort operation when seven aircraft bombed Lae, Salamaua and shipping in the target area. Despite the loss of one bomber and severe damage to another, the mission was a success and resulted in the sinking of one Japanese merchant vessel.

Combat operations commenced on June 25 1942 when attacks against shipping and Salamaua airfield were made. While this first mission was not especially successful, further attacks against Japanese positions were quite productive. Reconnaissance missions were also conducted and on October 6, the first torpedo strike – an unsuccessful attack by six Beauforts against Japanese naval forces – was flown. These low level operations, fraught with danger for the relatively slow and vulnerable Beauforts, met with little success and over a period of time a few ships – significantly including a cruiser off Gasmata – were damaged or sunk.

100 Squadron played a small but unsuccessful part in the Battle of the Bismarck Sea when on March 3 1943 eight torpedo armed Beauforts were dispatch against the by now scattered convoy. Unfortunately, only two aircraft found the ships and neither secured a torpedo hit. This unsuccessful mission proved to be 100 Squadron's last torpedo bombing mission and thereafter it operated solely in the level bombing mode.

As time went by, 100 Squadron found itself conducting most of its attacks against land targets throughout New Guinea. Perhaps the most important task allocated to the unit during the New Guinea campaign was the neutralisation of the Japanese fortress at Rabaul. This task was accomplished by large scale night bombing in conjunction with other Beaufort squadrons. These nocturnal operations in turn, supported large massed American daylight attacks on the stronghold.

Gasmata, amongst other targets, also came under attack by 100 Squadron but its Japanese defenders were able to inflict some losses on their Australian attackers. This was tragically highlighted on October 4 1943 when, of 10 Beauforts making a low level attack against the airfield, three were shot down. One of these Beauforts, captained by Flight Lieutenant R H Woolacott, continued its bombing run even after it was hit and had burst into flames. Holding his crippled aircraft steady, the pilot gave his bomb aimer sufficient time to release his bombs – all of which were direct hits in the centre of the runway – before the Beaufort crashed and exploded in the target area. None of the crew escaped.

In October 1943, 100 Squadron relocated to Goodenough Island and succeeding moves took the unit to Nadzab and Tadji. From these locations the unit continued its bombing operations well into 1945. At times the Beauforts gave direct support to Australian troops and in these operations made very accurate and damaging attacks against Japanese forces, especially in the drive to Wewak.

After the war 100 Squadron remained in New Guinea, finally disbanding at Finschafen on August 19 1946.

Formed: *February 15 1942*
Squadron Code: *QH*
Aircraft: *Beaufort*
Locations: *Richmond: February – May 1942*
Mareeba: May – July 1942
Laverton: July – October 1942
Bohle River: October 1942
Milne Bay: October – November 1942
Goodenough Island: November 1943 – May 1944
Nadzab: May – June 1944
Tadji: June 1944 – January 1946
Finschafen: January – August 1946
Disbanded: *August 19 1946*

**An early production Beaufort of 100 Squadron.
(RAAF Museum)**

102 SQUADRON

102 Squadron formed at Cecil Plains on May 31 1945 under the command of Squadron Leader J E Dennett. Allocated a heavy bomber role, the unit's first Liberators arrived in July. Training commenced in earnest, however, the war ended before 102 Squadron could become operational.

On August 16 nine of 102 Squadrons Liberators took part in a massed fly-past over Brisbane to celebrate VP Day. The Liberators were then used to ferry personnel back from the Pacific islands, particularly Bougainville and Moratai. Flying tapered off in December 1945 and 102 Squadron disbanded on March 19 1946.

Formed: *May 31 1945*
Squadron Code: *BV*
Aircraft: *Liberator*
Locations: *Cecil Plains: May 1945 – Mar 1946*
Disbanded: *March 19 1946*

(right) 102 Squadron flew Liberators.

107 SQUADRON

107 Squadron formed at RAAF Base Rathmines on May 10 1943. Under the temporary command of Flying Officer N J Lennon, the new Squadron was equipped with Vought-Sikorsky OS2U-3 Kingfisher float planes and personnel drawn from 3 Operational Training unit.

Anti-submarine patrols, convoy escorts and maritime patrol flights commenced immediately, however, no sightings of Japanese submarines, which had been operating off Australia's east coast, were made. 107 Squadron moved to St Georges Basin in mid 1944 where routine patrols continued.

When an American ship was sunk by a German submarine south of Jervis Bay on December 25 1944 patrols were intensified. The next day a Kingfisher captained by Flight Lieutenant O'Reilly sighted the submarine but was unable to attack as the aircraft was on a test flight and was unarmed. On December 29 Warrant Officer H T Moore located a possible periscope in the vicinity of the sinking and carried out a depth charge attack. A large quantity of oil came to the surface, however, as it was later learned that the submarine had already left the area, the source of the oil remains a mystery. This singularly unsuccessful event was destined to be the RAAF's only Kingfisher attack of the war.

With the arrival of the British Eastern Fleet in Australian waters in early 1945, 107 Squadron provided air sea rescue support to the Royal Navy. Several searches for lost aircraft were made and on one occasion a Kingfisher flown by Squadron Leader T Egerton rescued a Corsair pilot whose aircraft had crashed into the sea. Flying activities were scaled down during the latter part of 1945 and 107 Squadron disbanded on October 31 1945.

Formed: *May 10 1943*
Squadron Code: *JE*
Aircraft: *Kingfisher*
Locations: *Rathmines: May 1943 – July 1944*
St Georges Basin: July 1944 – October 1945
Disbanded: *October 31 1945*

A 107 Squadron Kingfisher. Despite the aircraft's small size the type had excellent endurance and conducted a considerable number of convoy escort and anti-submarine patrols off Australia's East Coast. (RAAF Museum)

450 SQUADRON

450 Squadron, the first RAAF unit activated under the terms of the Empire Air Training Scheme, formed at Williamtown on February 7 1941.

Without pilots or aircraft, the unit sailed for the Middle East on April 11 and, on arriving, was met by its Commanding Officer, Squadron Leader G Steege. Based at Abu Suier and still without pilots, 450 Squadron was combined with 260 Squadron, an RAF Hawker Hurricane fighter unit which, as yet, was without its ground staff. The combined squadron participated in operations against the Vichy French in Syria. Airfields, motor transport other targets being attacked up until the Vichy surrender in early July.

After the Syrian campaign, from July, 260/450 Squadron operated Hurricane flights at Beirut and Haifa. Soon afterwards 260 Squadron's ground staff arrived, the two squadrons were separated and 450 Squadron moved to Rayak where it was allocated Hurricane Is and Miles Magisters and acted as a temporary Operational Training Unit. After only two weeks, however, 450 Squadron shed its trainee pilots and aircraft and moved to Burg-El-Arab in October where its ground staff worked as an aircraft repair unit.

Finally, in January 1942, 450 Squadron received Curtiss Kittyhawk Is and commenced training as an operational fighter squadron. After a series of moves the unit flew its first operational mission – a fighter sweep west of Tobruk – on February 19. Three days later Sergeant R Shaw claimed 450 Squadron's first kill when he shot down a Ju 88.

450 Squadron operated intensively, bombing and strafing ground targets, flying escort missions and engaging in air combat with German and Italian fighters, many of which possessed better performance than the Australian aircraft. Despite what at times were heavy losses, 450 Squadron maintained its damaging attacks against the Axis forces and in the process earned for itself the rather grim nickname "The Desert Harassers".

In June 450 Squadron was forced to evacuate Gambut in the face of advancing tank columns. Moving initially to LG 75, as the retreat continued several further withdrawals were effected. Some of the airfields oc-cupied in this period were also subject to damaging attacked by Axis aircraft resulting in casualties and the loss of aircraft. On the night of August 2, for instance, 20 bombs fell in 450 Squadron's dispersal area which set one Kittyhawk on fire but, fortunately, did not cause any casualties. On some nights enemy bombers made several attacks – allowing little time for the Australians to sleep after a day's operations.

During the lead up to the Battle of El Alamein, 450 Squadron flew from LG 91 and operated intensively, particularly in the ground attack and air combat roles. After completing one such mission on October 25, a large sandstorm and shortage of fuel forced the Kittyhawks to divert to Maryut where they put down on one runway while a squadron of Hurricanes landed on the intersecting runway. Fortunately, there were no collisions as the fighters passed each other at the runway intersection. During the advance which followed the Allied breakthrough at El Alamein, 450 Squadron moved forward constantly while supporting the 8th Army's drive.

Several more moves bought 450 Squadron to Marble Arch where operations continued despite the airfield being heavily mined. Three ground staff were killed, four more were wounded and a taxying P-40 was blown up before the area was made safe by mine clearance personnel, who themselves suffered 20 fatalities in the process.

A 450 Squadron Kittyhawk is "bombed up" prior to a ground attack mission in Italy. (RAAF Museum)

The African campaign ended with the surrender of German and Italian forces on May 13 1943. In July 450 Squadron moved to Malta and then to Sicily when that island was invaded by the Allies. After the subjugation of Sicily, 450 Squadron participated in the Italian campaign, however, enemy aircraft were rarely encountered and flying was mostly confined to ground attack operations for the remainder of the war. The surrender of Italian forces on September 8 did not result in an end to the fighting and 450 Squadron moved forward several times as the Germans slowly gave ground up the Italian peninsula.

1944 saw 450 Squadron still engaged in the Italian campaign but also flying in support of Yugoslav partisans and attacking coastal shipping, with considerable success. As time went by the Germans developed an excellent anti-aircraft defence system which shot down a considerable number of Kittyhawks. Some pilots were bought down several times and losses would have been even heavier were it not for the P-40's rugged construction which allowed many to return despite severe battle damage.

On March 21 1945, 450 Squadron participated in a massed fighter/bomber attack against Venice Harbour. Results were very satisfying and included the sinking of two merchant ships and one motor torpedo boat, the destruction of five warehouses and major damage to a quay.

A most unusual incident occurred on April 20 when Flight Sergeant Eaves was shot down by his own ricocheting bullets while attacking a steel girder bridge. Fortunately, the pilot survived this extraordinary incident. 450 Squadron, which had begun to re-equip with Mustang IIIs, was operating from Cervia when the war in Europe ended and, after relocating again, disbanded at Lavariano on August 20 1945.

Formed: *February 7 1941*
Squadron Code: *OK*
Aircraft: *Hurricane, Magister, Kittyhawk, Mustang*
Locations: *Williamtown: Feb – April 1941*
Abu Sueir: May – June 1941
Amman: June – July 1941
Damascus: July 1941
Haifa: July – August 1941
El Bassa: August 1941
Rayak: August – October 1941
Damascus: October – November 1941
Burg el Arab: November – December 1941
Qassasin: December 1941 – January 1942

LG "Y": January 1942
Sidi Heneish: January – February 1942
Gambut: February – June 1942
Sidi Azeiz: June 1942
LG 75: June 1942
Sidi Heneish: June 1942
El Dada: June 1942
LG 106: June 1942
LG 91: June – October 1942
LG 224: October 1942
LG 175: October – November 1942
LG 106: November 1942
LG 101 and LG 76: November 1942
LG 142: November 1942
Gazala: November 1942
Martuba: November – December 1942
Belandah: December 1942
Marble Arch: December 1942
Amel-el-Chel: December 1942 – January 1943
Hamraret l: January 1943
Sedada: January 1943
Bir Dufan: January 1943
Castel Benito: January – February 1943
El Assa: February – March 1943
Neffatia: March 1943
Medenine: March – April 1943
El Hamma: April 1943
El Djem: April 1943
Kairouan: April – May 1943
Zuara: May – July 1943
Luga: July 1943
Pachino: July – August 1943
Agnone: August – September 1943
Grottaghe: September 1943
Bart: September 1943
Foggia: September – October 1943
Mileni: October – December 1943
Cuttalla: December 1943 – May 1944
San Angelo: May – June 1944
Guidonia: June 1944
Falenum: June – July 1944
Crete: July – August 1944
Lesi: August – September 1944
Foiano: September 1944
Lesi: September – November 1944
Fano: November 1944 – February 1945
Cervia: February – May 1945
Lavariano: May – August 1945
Disbanded: *August 20 1945*

451 SQUADRON

451 Squadron, an army co-operation unit, formed at Bankstown on February 15 1941 and sailed for Egypt in May of the same year.

Equipped with Hawker Hurricane Is at Qasaba, the unit, commanded by Squadron Leader V A Pope, undertook artillery spotting and tactical reconnaissance duties against German and Italian forces.

In November 451 Squadron supported British troops during the "Crusader" offensive flying tactical and photo reconnaissances as well as directing artillery. During this campaign 451 Squadron played an important role in locating enemy troop concentrations and fortifications whilst, on one occasion during the attack on Bardia in November 1941, an enemy ship was sunk in that port by artillery fire directed by an Australian Hurricane. Additionally a Hurricane detachment operated from within the Tobruk fortress conducting tactical reconnaissance missions in support of ground forces. During some of these missions the Hurricanes were intercepted by enemy fighters and several were lost.

After the conclusion of "Crusader", 451 Squadron was withdrawn from operational flying and placed on garrison duty in Syria with a detached flight on Cyprus. This was the beginning of a lengthy and extremely demoralising period for the unit. RAF Middle East Command at this time had more army co-operation squadrons than it could effectively utilise and proposed that 451 Squadron be redesignated a fighter squadron. For reasons unclear, the Australian Air Board strenuously resisted this action with the result that the RAF was unable to gainfully employ the unit.

451 Squadron remained almost totally inactive throughout 1942 and, as a result, morale plummeted to a dangerously low level. Leaders at all levels within the unit were unable to check this deterioration and eventually the Commanding Officer was relieved of his command. The situation with 451 was unique and no other RAAF squadron, before or since, has suffered such a widespread loss of morale and combat effectiveness.

Finally, in January 1943, 451 Squadron returned to Egypt and, at last redesignated a fighter squadron, was made responsible for providing the fighter defence of part of the Nile Delta. To assist in this task, in February, some Spitfire VCs were allocated for high altitude interception work. Unfortunately, enemy aircraft rarely ventured so far behind Allied lines, with the result that the unit still failed to find the action its personnel craved. Convoy escort patrols were invariably uneventful and, up until July, only one inconclusive aerial combat occurred.

In March 1944, 451 Squadron moved to Corsica and, re-equipped with Spitfire IXs, supported the invasion of southern France. All ranks looked forward eagerly to the coming fight, however, their hopes were again dashed as enemy opposition to the invasion was slight. Patrols over the bridgehead and bomber escort missions, much to the frustration of air and ground crews alike, were regularly uneventful.

451 Squadron's fortunes finally changed towards the end of 1944 when it moved to the United Kingdom. Initially operating from Hawkinge, active operations comprising ground attack missions, bomber escort tasks and fighter sweeps over enemy occupied territory were flown until Germany's surrender in May 1945.

After the close of hostilities, 451 Squadron moved to Germany as part of the Allied occupation force but plans to maintain a long term RAAF presence in Germany withered and 451 Squadron disbanded at Wunsdorf on January 21 1946.

Formed: *February 15 1941*
Squadron Codes: *BQ, NI*
Aircraft: *Hurricane, Spitfire*
Locations: *Bankstown: February – May 1941*
Kasarfeet: May – June 1941

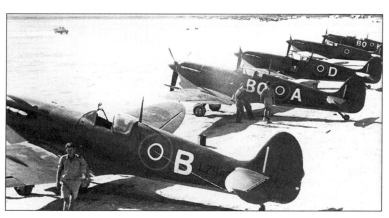

After operating Hurricanes and Kittyhawks, 451 Squadron re-equipped with Spitfires. (RAAF Museum)

Qasaba: June – October 1941
LG 75: October – November 1941
LG 132: November 1941
Sidi Azeiz: November – December 1941
Maddalena: December 1941
Sidi Azeiz: December 1941 – January 1942
Heliopolis: January 1942 – February 1942
Rayak: February – November 1942
St Jean: November 1942 – January 1943
Mersa Matruh: January – February 1943
Edku: February – August 1943
El Daba: August 1943 – February 1944
Gamil: February 1944
Almaza: February – April 1944
Poretta: April – May 1944
Serragia: May – July 1944

St Catherine: July – August 1944
Cuers: August – October 1944
Foggia: October – November 1944
Hawkinge: December 1944 – February 1945
Manston: February 1945
Swannington: February – April 1945
Lympne: April – May 1945
Hawkinge: May 1945
Skaebrae: May – September 1945
Hornchurch: September 1945
Fassberg: September 1945
Wunsdorf: September – November 1945
Gatow: November – December 1945
Wunsdorf: December 1945 – January 1946
Disbanded: January 21 1946

452 SQUADRON

The first RAAF unit to form in the United Kingdom, 452 Squadron came into existence at Kirton-in-Lindsay on April 8 1941.

Commanded by the RAF's Squadron Leader R G Dutton, a Battle of Britain ace with 19 enemy aircraft to his credit, the new fighter squadron was equipped with Supermarine Spitfire Is.

Destined to become remarkably successful in a short time, 452 Squadron achieved operational status on June 2 and combat flying began immediately. Convoy patrols and other defensive tasks were later supplemented by bomber escort missions and fighter sweeps over occupied France and it was during one of these operations on July 11 that the unit obtained its first kill when Flight Lieutenant B E Finucane shot down a Bf 109. Unfortunately, one Australian pilot, Sergeant Roberts, failed to return from this mission.

On July 21, 452 Squadron moved to Kenley where its operational flying intensified. The unit found itself constantly in action with German aircraft – particularly Messerschmitt Bf 109 fighters – and in the large scale combats which resulted for the remainder of the year, 452 Squadron soon became the leading fighter squadron in Fighter Command. In July alone, 22 Bf 109s were claimed destroyed with the top scoring pilot being "Paddy" Finucane whose share was nine aircraft.

Several of 452 Squadron's junior Australian pilots were also achieving good results and these included Sergeant K B Chisholm, and Flying Officer R E Thorold-Smith. Pilot Officer Keith Truscott, an officer destined to become the RAAFs second highest scoring ace of World War II, also obtained his first victories in this period.

During one combat in September, 452 Squadron was involved in a bitter engagement in which four pilots were shot down and killed. Despite these losses, the unit came off the better destroying six Bf 109s, probably destroying another and damaging a further two. September also saw 452 Squadron claim its first Focke Wulf Fw 190 – the newly promoted Flight Lieutenant Thorold-Smith being the successful pilot. Several more of these superior German fighters later fell to the guns of the Australian pilots.

Operations from Kenley as well as Redhill continued for the remainder of 1941 and into the new year. In February 1942, 452 Squadron was involved in the abortive attempts to halt the "Channel Dash" by major surface units of the German Navy. Flying in deplorable weather conditions, the Spitfires located some of the warships and damaged a destroyer with their cannon and machine gun fire.

February also saw a remarkable escape from impossible odds when Sergeant P Makin became separated from his formation near Cape Gris Nez. Attacked by 12 Bf 109s, the lone Australian nonetheless managed to damage one enemy fighter before making a safe return to Kenley.

On March 13, Squadron Leader Truscott, by now an ace with an impressive tally of kills to his credit, claimed 452 Squadron's last victory in the European theatre when he destroyed an Fw 190. 452 Squadron then moved to Andreas on the Isle of Man where it remained for three months.

During its time in the UK, 452 Squadron had, for the loss of 22 pilots, been credited with destroying 62 German aircraft, probably destroyed another seven and damaged a further 17.

In June the personnel of this battle seasoned and highly regarded fighter squadron sailed for

Australia – a country in desperate need of fighter aircraft to protect its vulnerable north from Japanese attack. Disembarking in Melbourne, the unit's personnel moved to Richmond where they waited a considerable time for Spitfire VCs, forwarded from the UK, to arrive. Once these were received and some refresher training had been conducted the unit departed for Darwin in January 1943.

Established at Strauss in the latter part of January, 452 Squadron had a short period to settle into its new area of operations before the first Japanese attack after the Spitfires' arrival occurred on March 15. In the ensuing battle, which involved some 40 Japanese fighters and bombers, Flying Officer A P Goldsmith destroyed a bomber and a fighter however Squadron Leader Thorold-Smith was shot down and killed.

After a lull during April, Japanese air attacks recommenced in May. On May 2 Darwin's radar detected a force of over 20 bombers with a strong fighter escort approaching. Along with the other Spitfire squadrons charged with Darwin's defence, 452 Squadron scrambled and engaged the Japanese formation after they had bombed their targets. During the ensuing combat one bomber and two fighters were downed, two fighters probably destroyed and two fighters damaged. Unfortunately, three of the 11 Spitfires were shot down and one force landed. Two pilots were killed.

On June 20 Squadron Leader R S MacDonald led the Spitfires into action again when the unit intercepted a large Japanese attack. In their initial pass the Spitfires destroyed three bombers and one fighter before the sky filled with twisting and diving aircraft. This success was marred somewhat by the loss of Pilot Officers W E Nichterlein and A T Ruskin-Rowe who were both shot down and killed.

After an abortive interception eight days later, 452 Squadron was involved in an unfortunate incident while intercepting a Japanese attack on June 30. Misunderstanding their orders, the pilots dived on the bombers instead of their escort of Zero fighters with the result that the Spitfires themselves came under a concerted attack from the fighters. Three of the 12 Spitfires scrambled were lost (including one which crashed on takeoff). One of the luckiest Australian pilots was Flight Sergeant C R Duncan who, after baling out, spent five gruelling days in rough country before

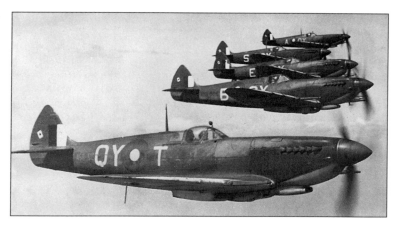

452 Squadron Spitfires in formation after the unit was sent to Australia. (RAAF Museum)

being rescued. Two enemy aircraft were destroyed and the unit shared in the destruction of another.

Further battles over Darwin were fought over the succeeding months and, despite operating an aircraft not suited to the hot and humid tropical conditions of the Northern Territory, the Australian pilots proved themselves a match for their Japanese counterparts who, for their part, had to continually accept heavy losses during their attacks on Darwin. From July, Japanese air activity over the Northern Territory had become spasmodic with the result that for weeks at a time the Spitfires had little to do.

A detachment sent to Millingimbi during August with the purpose of disrupting Japanese floatplane activity out to sea from this base had some success on August 10 when one floatplane was shot down and another damaged. August also saw the Japanese commence night attacks in the Darwin area and in an unsuccessful attempt to counter these nocturnal visits, 452 Squadron sent up some aircraft but failed to make contact with the raiders.

September saw a slight resurgence of air activity when the Japanese, tiring of having their unescorted reconnaissance aircraft shot down over Darwin, began to provide heavy fighter escorts for these machines. In the combats which resulted several Japanese aircraft were claimed.

By the end of the year Japanese air activity was almost non existent and little useful work remained for the unit. This unfortunate state of affairs continued until January 1945 when 452 Squadron moved to Moratai. From here the Spitfires conducted ground attack operations against bypassed Japanese island garrisons until the end of the war. After supporting the invasions of Tarakan and Balikpapan, 452 Squadron disbanded on November 17 1945.

Formed: *April 8 1941*
Squadron Codes: *UD, QY*
Aircraft: *Spitfire*
Locations: *Kirton-in-Lindsay: Apr – July 1942*
Kenley: July – October 1941
Redhill: October 1941 – January 1942
Kenley :January – March 1942
Andreas: March – June 1942

Richmond: September 1942 – January 1943
Strauss: January 1943 – March 1944
Pearce: March 1944
Strauss: March – July 1944
Sattler: July 1944 – January 1945
Moratai: January – June 1945
Tarakan: June – November 1945
Disbanded: *November 17 1945*

453 SQUADRON

453 Squadron formed at Bankstown on May 23 1941. Under the command of Squadron Leader W F Allshorn, the new squadron sailed for Singapore where it was to be supplied with aircraft and equipment from RAF sources.

By August 453 Squadron was established at Sembawang and equipped with Brewster Buffalo Mk.Is. These obsolete American built fighters had proved unsuitable for use in Europe and as cast offs were sent to the Far East in considerable numbers.

With Japan's surprise attack on Malaya, 453 Squadron was bought to a state of readiness and on December 13 deployed its 16 aircraft to Ipoh to support Allied troops already retreating before the Japanese. Three aircraft were destroyed in accidents before reaching Ipoh, however, three others met with immediate success when scrambled to intercept unescorted bombers attacking Ipoh. Flying Officer R D Vanderfield destroyed two bombers while Sergeants M N Read and V A Collyer shared in the destruction of three dive bombers.

Ground support operations were flown later that day but in the afternoon 40 Japanese fighters arrived over the base, three of which were shot down by Flight Lieutenant T A Vigors before he and another pilot met the same fate (both were wounded but survived after baling out). Unfortunately, several aircraft were destroyed on the ground in this attack. Over the succeeding days, many operations were mounted despite continued attacks on Ipoh which caused further losses.

After a few days 453 Squadron moved to Kuala Lumpur where the unit – now the only operational fighter squadron left in Malaya – received replacement aircraft. The Japanese quickly targeted this base, however, and over a period of several days more Buffalos were destroyed.

On December 22, the heaviest attack yet mounted on Kuala Lumpur commenced and resulted the destruction of five Buffalos, damage to four more and the deaths of three pilots. In return, the vastly outnumbered RAAF Squadron claimed the destruction of four Japanese aircraft. Army personnel subsequently found the wreckage of 10 aircraft in the jungle.

With only three serviceable aircraft remaining, 453 Squadron was withdrawn to Sembawang where on December 24, it merged with 21 Squadron and became known as 21/453 Squadron. The combined unit was allocated 16 Buffalos and attempted to defend Singapore which was now under regular attack by Japanese aircraft. Where possible, tactical reconnaissance and ground attack missions, in support of ground troops still fighting on the mainland, were also flown. Continued losses, both in the air and on the ground, combined with a lack of replacement aircraft, ensued that the combined RAAF squadron quickly dwindled in numbers and effectiveness.

On January 26 453 and 21 Squadrons were separated – the latter almost immediately sailed for the Netherlands East Indies. 453 Squadron remained at Singapore operating the last six Buffalos available but, due to the aircraft's low speed and poor rate of climb, little could be done to interfere with continuing Japanese attacks.

In the first days of February 1942 the personnel and four surviving aircraft were evacuated to Java just days before Singapore fell. With no replacement aircraft and spares available, 453 Squadron could not be made operational and consequently its personnel sailed for Australia and the unit formally disbanded en route to Adelaide.

453 Squadron reformed at Drem in England on June 18 1942. Equipped with Spitfire VBs, the unit moved to Hornchurch on September 14 where it met enemy aircraft for the first time on October 31. In this combat three fighters and a bomber were damaged for the loss of one Spitfire. Moving to Southend early in December, fighter sweeps, bomber escorts and occasional attack missions were flown and the unit's score of enemy aircraft slowly mounted.

On June 30 1943, 453 Squadron relocated to Ibsley before moving again on August 18 to

Perranporth. On October 8 a formation of Spitfires led by Flight Lieutenant D G Andrews located eight Bf 110s low over the sea near Brest. The Spitfires dived on their prey and destroyed five of the enemy aircraft for the loss of two of their own number (the pilot of one, Flying Officer R H S Ewins was later rescued). Three days later, the unit moved again, this time to Skeabrae.

Early 1944 saw 453 Squadron training in the ground support role, while at the same time maintaining defensive patrols designed to keep German reconnaissance aircraft from detecting the shipping build up prior to the invasion of Europe.

On May 27 the first of many dive bombing missions against targets in occupied France was flown while on D Day itself, and for some time to follow, patrols over the beachhead were maintained. Generally these were of short duration until the Spitfires began to stage through French landing grounds, which enabled them to remain overhead for longer periods. This was a particularly exciting period for the Australian pilots and the unit flew intensively, completing armed reconnaissances, fighter sweeps, bomber escorts, shipping sweeps and patrols. On one forward airfield 453 Squadron even came under artillery fire!

On June 16 a formation of 12 Bf 109s were engaged over Caen and two of these were destroyed and another two probably destroyed. By the end of the month, 453 Squadron was based at Longues while in September it moved forward several times behind the advancing Allied ground forces.

The end of September saw a withdrawal to Coltishall and then Matlaske in England where the Spitfires were used against German V-1 and V-2 missile launching sites and depots in Holland. This activity entailed a considerable flying effort and losses, to anti-aircraft fire, were frequent.

453 Squadron moved to Lympne in April and on May 2 to Hawkinge where ground attack and bomber escort missions continued until Germany capitulated. 453 Squadron was selected as part of the Allied occupation forces and moved to Fassberg in Germany during August. Two weeks later it moved to Gatow where it became the first Commonwealth squadron based in the German capital. However, this stay was a short one and on October 18 453 Squadron moved a final time to Wunsdorf where it disbanded on January 21 1946.

Formed: *July 29 1941*
Squadron Codes: *TD, FU*
Aircraft: *Buffalo, Spitfire*
Locations: *Bankstown: July – August 1941*
Sembawang: August – December 1941
Ipoh: December 1941
Kuala Lumpur: December 1941
Sembawang: December 1941 – February 1942
Batavia: February 1942
Disbanded:
Drem: June – September 1942
Hornchurch: September – November 1942
Martlesham Heath: November – December 1942
Rochford: December 1942 – March 1943
Hornchurch: March – June 1943
Ibsley: June – August 1943
Perranporth: August – October 1943
Skeabrae: October 1943 – January 1944
Detling: January – April 1944
Ford: April – June 1944
Longues: June – September 1944
Beauvais: September 1944
Douai: September 1944
Deurne: September 1944
Coltishall: September – October 1944
Matlaske: October 1944 – April 1945
Lympne: April – May 1945
Hawkinge: May 1945
Lasham: June – September 1945
Fasberg: September 1945
Gatow: September – October 1945
Wunsdorf: October 1945 – January 1946
Disbanded: *May 31 1945*

Buffalo pilots of 453 Squadron with one of the unit's aircraft. (RAAF Museum)

454 SQUADRON

Formed at Williamtown on May 23 1941, 454 Squadron – which at that stage comprised of ground staff only – sailed for the Middle East where it was to join its aircrews and aircraft.

These apparently simple plans did not come to immediate fruition, however, and on arrival the Australians were used to service Halifax and Liberator aircraft. This misemployment continued for some months and it was not until September 1942 that 454 Squadron was bought together at Aqir in Palestine where it received its own aircraft, Bristol Blenheim Vs.

Ostensibly a light bomber unit, 454 Squadron's first task had little to do with its nominated role as the unit was used to provide refresher training for Blenheim aircrews prior to their being posted to RAF squadrons operating the same type.

Finally, after moving to Gianaclis in January 1943, 454 Squadron began active operations when its role was changed to that of a general reconnaissance unit and it converted to Martin Baltimore IIIs. In its new capacity anti-submarine patrols, anti-shipping strikes, leaflet dropping and, at times, bombing attacks – principally against targets in Greece and Crete – were carried out.

Unfortunately, after being bought together in such a haphazard manner, the unit was filled with additional personnel with scant regard to its national identity with the result that very few of the aircrew were Australians and, notwithstanding the draft of groundstaff sent from Williamtown, a significant portion of the tradesmen were of other nationalities.

The anti-submarine patrols in particular were usually uneventful and little was seen. On one occasion, shortly after converting to its new aircraft however, a Baltimore captained by Flying Officer Lewis was attacked by two Bf 109 fighters.

In a surprising turn of events the lumbering bomber's gunner, Flying Officer Carruthers, shot down one fighter and so badly damaged the other that it withdrew. The Baltimore returned safely to base. In a number of similar instances Baltimore crews bested their better armed attackers.

These successful engagements were later marred by 454 Squadron's most disastrous operation of the war. During a specially requested low level attack against Creten factories, barracks and road targets on July 23 1943, six out of eight Baltimores despatched were shot down by anti-aircraft fire while both surviving aircraft were damaged.

Anti-submarine and general reconnaissance patrols continued as the unit's main activity throughout 1943 and into 1944. Several aircraft disappeared during these missions, however, of the enemy shipping located, a significant portion was sunk by attack aircraft alerted by the patrolling Baltimores.

454 Squadron itself participated in a major anti-shipping strike on July 1 1944. A convoy of three merchant vessels escorted by ten naval vessels was initially located by a patrolling RAF aircraft. Relays of RAAF Baltimores were sent out to shadow the convoy – despite a fighter escort which harried them and shot one down – while a strike force was gathered. The combined formation, which included three 454 Squadron aircraft, attacked through heavy anti-aircraft fire sinking or damaging several ships. 454 Squadron's share was a direct hit on a 2000 tonne merchantman which was later confirmed as sunk.

454 Squadron moved to Italy in July 1944 and from a series of new bases, flew intensively in the daylight bomber role striking targets throughout Italy and Yugoslavia. From January 1945 the Baltimore crews retrained for a new role – that

454 Squadron Baltimores in the Middle East. (RAAF Museum)

of night intruder work. In this phase, the unit's losses were not unduly heavy although enemy defences could not be entirely discounted. A number of aircraft were lost and in one particular incident, Warrant Officer Hogan bought his Baltimore back to base holed in 150 places, with its starboard engine windmilling uselessly and the entire crew wounded. 454 Squadron flew its last operational flights – 14 armed reconnaissances – on May 1 and disbanded at Villaorba on August 14 1945.

Formed: *May 23 1941*
Squadron Code: *Nil*
Aircraft: *Blenheim, Baltimore*

Locations: *Williamtown: May 1941 –*
Aqir: September – October 1942
Oaiyara: October 1942 – January 1943
Gianaclis: January – February 1943
LG 91: February – April 1943
Gambut: April – August 1943
LG 91: August – October 1943
St Jean: October – November 1943
Berka 111: November 1943 – July 1944
Pescara: July – August 1944
Falconara: August – December 1944
Cesenatico: December 1944 – May 1945
Villaorba: May – August 1945
Disbanded: *August 14 1945*

---

455 SQUADRON

Destined to be the first RAAF unit in RAF Bomber Command, 455 Squadron formed at Williamtown on May 23 1941.

The airmen at Williamtown sailed for England the following month where they joined others gathered at Swinderby to complete the unit's personnel establishment. The Squadron, equipped with Handley Page Hampden I medium bombers and commanded by Wing Commander J Gyle-Murray, commenced operations on August 29 when a solitary bomber captained by Squadron Leader D French, joined in a night raid to Frankfurt.

Operations quickly increased in intensity and a considerable number of mine laying attacks were mounted against the lightly defended French coastal areas. Later, as the unit concentrated more on bombing attacks against German industrial targets, losses from anti-aircraft and night fighter defences increased.

Early in February 1942, 455 Squadron relocated to Wigsley and, within days, participated in an unsuccessful attack against the German battle cruisers *Scharnhorst, Gneisenau* and the cruiser *Prince Eugen*. In April the unit transferred to RAF Coastal Command and retrained in the torpedo bombing role – even though the by now obsolete Hampdens were quite unsuited to this role. Notwithstanding their obsolescence, the outdated Australian aircraft maintained a high rate of effort and over time managed to sink or damage a number of enemy ships.

Most of 455 Squadron deployed to the Soviet Union during September to support the movement of Russian bound convoys which at that time were suffering grievous losses. Three of the 16 Hampdens were lost prior to arriving at their new base and the remainder, after completing only one precautionary anti-shipping sweep, were handed over to the Soviets who were then instructed in their use by the RAAF personnel. The detachment's purpose had been successfully fulfilled, though while the Australians were at Murmansk, the expected movement of German warships from Norwegian ports did not eventuate.

By the end of October 455 Squadron was reunited at Sumburgh and allocated replacement Hampdens. While the main body of the Squadron had been in the USSR, the remaining aircraft carried out anti-shipping and anti-submarine patrols – a task which would occupy the reunited squadron for the remainder of 1942 and into 1943. These flights were usually uneventful, however, two merchant ships (one of 3570 and the other of 6120 tonnes) were sunk, while on April 30 1943 Flight Sergeant J S Freeth's crew destroyed a German submarine (U-227) north of the Shetland Islands.

An impressive maritime success occurred on January 28 1943 when seven Hampdens from 455 and 487 Squadrons secured no less than six torpedo hits on a 3570 tonne ship which quickly sank. On May 12 a 6018 tonne ship, located by Squadron Leader J N Davenport and Flying Officer B Atkinson, was sent to the bottom near Egero Island after one torpedo hit vessel. The vulnerable Hampdens were occasionally intercepted whilst on their shipping sweeps although one Ju 88 pilot's glee at locating four such easy targets on February 21 1943 quickly turned to dismay when his aircraft was so badly damaged by return fire that he had to break off his attack.

On October 11 1943, 455 Squadron moved to Leuchars and re-equipped with more capable aircraft – hard hitting Bristol Beaufighter Xs. After converting to these deadly aircraft, the unit moved to Langham, where it and 489 Squadron formed one of Coastal Command's new strike wings.

Operating in the flak suppression role for the torpedo carrying Beaufighters of 489 Squadron, operations commenced on May 6 1944 when several enemy vessels were attacked, and one sunk. In this new role the Beaufighters faced the hazards of heavy naval anti-aircraft barrages and, when operating in Norwegian waters especially, fighter opposition. Additionally, the Beaufighters often had to attack their targets up very narrow and precipitously walled Norwegian fiords – a task involving extreme hazard. Consequently Beaufighter losses, at times, were heavy indeed.

A typical operation in Norwegian waters occurred on November 8 when six Australian Beaufighters joined in an attack on enemy shipping located in Midgulen Fiord. Despite anti-aircraft fire and extremely hazardous flying conditions, two enemy ships were sent to the bottom.

Combined anti-shipping strikes, which resulted in the destruction of a number of vessels, continued into 1945 before the unit moved to Thornaby, where the Beaufighter's struck at German shipping in the Baltic Sea during the last days of the war. The Australian Beaufighters flew their last successful operation on May 3 when they left two minesweepers blazing while accom-

Although declared obsolete early in the war, 455 Squadron operated Hampdens for a considerable time initially in the night bombing and subsequently in the anti shipping role. (RAAF Museum)

panying New Zealand aircraft torpedoed a tanker. 455 Squadron barely survived Germany's surrender, disbanding at Thornaby on May 25 1945.

Formed: *May 23 1941*
Squadron Code: *UB*
Aircraft: *Hampden, Beaufighter*
Locations: *Williamtown: May – June 1941*
Swinderby: June 1941 – February 1942
Wigsley: February – April 1942
Leuchars: April 1942 – April 1944
Langham: April – October 1944
Dallachy: October 1944 – May 1945
Thornaby: May 1945
Disbanded: *May 25 1945*

---

456 SQUADRON

456 Squadron formed at Valley in the United Kingdom on June 30 1941.

Equipped with Boulton Paul Defiant Mk.I two seat night fighters, the unit was commanded by Squadron Leader G Olive. The first uneventful missions were flown on September 4 but from the end of the month radar equipped Bristol Beaufighter Mk.IIs began to replace the Defiants although both types operated together for some time.

On the night of January 10/11 1942 a Beaufighter crewed by Squadron Leader Hamilton and Pilot Officer Norris-Smith claimed 456 Squadron's first kill when they shot down a German bomber. Apart from this incident the unit saw little action, its time being spent on uneventful patrols, training and occasional air-sea rescue tasks, plus re-equipping with Beaufighter VIs.

Two more enemy aircraft were destroyed by the time the unit again began to re-equip, with

de Havilland Mosquito Mk.IIs, in December and commenced offensive missions, known as 'Rangers', over occupied Europe. These operations met with some success against ground targets. In March 1943, 456 Squadron moved to Middle Wallop, a more active area for night fighter operations. During periods of bad weather the Mosquitos were also utilised in the long range day fighter role.

On May 6 1943, Flying Officer P Panitz and his navigator, Pilot Officer R Williams, flew a especially successful 'Ranger' mission in which they damaged or destroyed six locomotives. These two airmen achieved considerable fame whilst serving in 456 Squadron and of 18 trains damaged or destroyed in this month, no less than ten were claimed by this crew.

In mid 1943, 456 Squadron made its first detachment to Predannick where the Mosquitos hunted German Ju 88 'fighters' which themselves were seeking vulnerable Coastal Command aircraft. Just as the Coastal Command aircraft were easy prey for the German airmen, so too were the Ju 88s no match for the Mosquitos. When located, the German formations usually scattered but were invariably caught and suffered losses.

Having spent short periods at Colerne and Fairwood Common, on February 29 1944, 456 Squadron again moved, this time to Ford, where, re-equipped with Mosquito's possessing more effective radar, it operated against bombers making night attacks against London. Victories quickly mounted in the period leading up to the invasion of Europe – one of the most successful crews being Wing Commander K Hampshire and Flying Officer T Condon. On the night of March 27/28, these two airmen destroyed two of the three bombers bought down over the United Kingdom by 456 Squadron that night.

In June the Mosquitos covered the Allied invasion of Europe and on the night of June 6/7, Flying Officer R G Pratt shot down a Heinkel He 177 near France – the first enemy aircraft destroyed since the landing. Although German air activity against shipping and the bridgehead was limited, 456 Squadron had outstanding success over France and shot down 13 bombers in June – mostly He 177s. The same month saw the commencement of V-1 flying bomb attacks against English cities and 456 Squadron destroyed a number of these pilotless aircraft. Flight Lieutenant K A Roediger claimed 456 Squadron's first V-1 and, with an eventual score of nine, he later became the highest scoring V-1 killer in the unit.

Late 1944 saw 456 Squadron move to Church Fenton where it commenced long range night at-

456 Squadrons "A Flight" aircrew at Colerne in November 1943. Those personnel in lighter uniforms are probably RAF. (RAAF Museum)

tack and bomber escort work. In this role the Mosquitos ranged all over Germany attacking airfields in support of Bomber Command night attacks as well as targets of opportunity.

456 Squadron again moved in March 1945, this time to Bradwell Bay where operations continued with German airfields being the main target. The war in Europe was nearly over, however, and the Mosquito's flew their last missions on May 2. Five days later Germany surrendered and, on June 15 1945, 456 Squadron disbanded.

Formed: *June 30 1941*
Squadron Codes: *PZ, SA, RX*
Aircraft: *Defiant, Beaufighter, Mosquito*
Locations: *Valley: June 1941 – March 1943*
Middle Wallop: March – October 1943
Colerne: October – November 1943
Fairwood Common: Nov 1943 – Feb 1944
Ford: February – December 1944
Church Fenton: December 1944 – March 1945
Bradwell Bay: March – June 1945
Disbanded: *June 15 1945*

---

457 SQUADRON

The RAAF's second United Kingdom based fighter squadron formed with Supermarine Spitfire Is at Baginton on June 16 1941 under the command of Squadron Leader P M Brothers.

The new unit's commander was a Battle of Britain ace with ten confirmed victories. Both he and his two flight commanders were RAF officers while the majority of ground staff were, initially, RAF personnel. These personnel were gradually

replaced by Australians and from the start, most pilots aside from those mentioned were Australian.

Declared operationally proficient in August, 457 Squadron moved to Jurby on the Isle of Man. Unfortunately, the routine patrols and convoy escort missions undertaken failed to result in any contact with the enemy throughout 1941 and into the first quarter of 1942. At one stage, 457 Squadron found itself being used as an operational training unit supplying Spitfire pilots to squadrons engaged in more active operations.

Finally, in March 1942, after moving to Redhill, 457 Squadron gained its first victory when New Zealand ace, Flight Lieutenant H L North, shot down an Fw 190 during an anti-shipping strike. Operations quickly increased in intensity and the unit found itself escorting light bomber attacks against targets in occupied France or making fighter sweeps over enemy territory. These operations were bitterly contested both by the Luftwaffe and anti-aircraft defences which immediately began to take a toll on the Australian pilots.

The German Focke Wulf Fw 190 fighters encountered proved to be a superior aircraft to the Spitfire Vs then being flown and in most combats more Spitfires were lost than enemy aircraft. A typical example of one of these unequal battles occurred on March 28 when 12 457 Squadron aircraft were attacked by between 40 and 50 Fw 190s near Cap Gris Nez. Pilot Officer R H C Sly destroyed one enemy fighter, Pilot Officer G G Russell probably destroyed another and Flight Lieutenant North damaged one more but in return three Spitfires were shot down. A further Spitfire was damaged and its wounded pilot, Sergeant W H Wright, only just managed to reach Redhill.

On April 29, 457 Squadron gained its last successes in Europe when Squadron Leader Brothers probably destroyed an Fw 190 and Pilot Officer D D MacLean damaged another. Fighter sweeps and escort missions continued into May but, although several more Spitfires were lost, no more victories were claimed before the unit was withdrawn from operations and moved to Kirton-in-Lindsay preparatory to being sent to Australia. During its short period of active operations 457 Squadron had destroyed five, probably destroyed four and damaged seven German aircraft.

In June 1942, 457 Squadron sailed for Australia arriving in September. Unfortunately, consignments of Spitfires sent afterwards were redirected to the Middle East with the result that the unit remained inactive at Richmond for longer than planned. Finally, aircraft were received and, after a period of refresher training, 457 Squadron moved to Livingstone in the Darwin area where, with the Spitfires of 54 (RAF) and 452 Squadrons, it became responsible for the air defence of Darwin which, at that stage, was under regular Japanese attack.

Well established at its new base by February 1943, 457 Squadron claimed its first victory of the Pacific war on March 7 when Flight Lieutenant D H McLean and Sergeant F R J McDowall shared in the destruction of a Dinah reconnaissance aircraft near Darwin.

Just over a week later, on March 15, 457 participated in a large scale combat when 21 bombers and 24 fighters attacked Darwin. In the ensuing melee the Spitfires destroyed one Betty bomber, two fighters and probably destroyed three more fighters.

A lull followed this combat and it was not until May 2 that another large Japanese attack against Darwin was made. In the ensuing combat, 457 Squadron claimed the destruction of one fighter with another probably destroyed and another two damaged. Two Spitfires were lost – the pilot of one being killed. It was quickly noted from experience gained in battles such as this one that the Spitfire's minimal endurance and high rate of engine wear from the prevailing dusty conditions would prove a recurring problem in the South West Pacific, particularly around Darwin where continuous losses were experienced from non operational causes such as engine failure and fuel exhaustion. Certainly the Spitfire never achieved the same degree of ascendancy over the Japanese as it did over the Germans in Europe.

Combat over Darwin continued and, while losses were sustained, some pilots began to take a considerable toll on the Japanese raiders. After the Japanese had made some attacks on Millingimbi, a detachment of Spitfires moved there on the morning of May 9 to counter further enemy activity. Later that same day Japanese fighters arrived overhead and two were shot down

Spitfires of 457 Squadron, probably in the Darwin area. (RAAF Museum)

for the loss one Spitfire which, during a low level chase, struck the ground, somersaulted several times and was totally destroyed. Remarkably, the aircraft's pilot, Pilot Officer B Little, escaped with only minor injuries! In a separate incident during the day, a detachment pilot shot down a Japanese floatplane into the sea.

457 Squadron participated in further engagements over Darwin, the last major battle occurring on July 6. One Zeke was bought down and five Bettys damaged but three pilots from 457 Squadron were also killed. From the end of July Japanese air activity over Darwin was severely curtailed and boredom became the unit's main enemy. The Japanese continued to send high flying Dinah reconnaissance aircraft over, however, and one of these was shot down over Fenton by Squadron Leader K E James on August 20. A second Dinah was destroyed later that day by two other pilots.

Tiring of their losses in unescorted reconnaissance aircraft, on September 7 a bomber with a large fighter escort appeared. In the ensuing fight a number of Spitfires from other squadrons were lost, although 457 Squadron itself conducted a very successful attack shooting down four Zekes and damaging others for no loss to itself.

Very little active flying occurred for the remainder of 1943 and in fact it was not until March 1944, when fear of a Japanese naval foray into the Indian Ocean forced the emergency deployment of 457 Squadron to Western Australia, that some excitement occurred. The Spitfires departed on March 8 and after a nightmare journey lasting several days, in which the pilots encountered severe tropical storms and dust clouds, they arrived at Guildford to find that the feared Japanese attack against Perth did not eventuate. Towards the end of the month the Spitfires returned to Livingstone.

On April 17 several Spitfires, staging through Bathurst Island, strafed barges, huts and a wireless station on Babar Island. All the aircraft returned from this, the unit's first ground attack operation from the Darwin area.

In May a second deployment to Western Australia occurred when the unit moved to Exmouth to provide fighter protection to the British Far Eastern Fleet which was using Exmouth as a temporary base. Uneventful patrols were flown before the Fleet moved on and 457 Squadron returned to the Northern Territory.

457 Squadron moved north to Moratai in early 1945 and flew its first operation – an armed reconnaissance over the Halmaheras during which an enemy aircraft was strafed on Galela airfield – on February 10. From this time on, the Spitfires were utilised in the ground attack role – a task not especially suited to these short range, minimally armed fighters.

The Spitfires later supported the invasion of Labuan and, from June, were based on that island. On June 20 two patrolling Spitfires flown by Flight Lieutenants J G B Campbell and S G Scrymgeour surprised and shot down a Dinah – the unit's first kill since 1943. 457 Squadron disbanded at Labuan on November 7 1945

Formed: *June 16 1941*
Squadron Codes: *BP, XB, ZP*
Aircraft: *Spitfire*
Locations: *Baginton: June – August 1941*
Jurby: August 1941 – March 1942
Redhill: March 1942 – May 1942
Kirton-in-Lindsay: May – June 1942
Richmond: September 1942 – January 1943
Livingstone: January 1943 – March 1944
Guildford: March 1944
Livingstone: March – February 1945
Moratai: February – June 1945
Labuan: June – November 1945
Disbanded: *November 7 1945*

---

458 SQUADRON

458 Squadron formed with only ground staff at Williamtown on July 8 1941 and proceeded to Britain in August to join aircrews and other personnel assembled at Holme-on-Spalding Moor.

Like most RAAF Empire Air Training Scheme squadrons, formed under such circumstances, 458 Squadron contained a significant number of RAF, RCAF and RNZAF personnel, and RAAF aircrew numbers were, initially, very low. Even the Commanding Officer, Wing Commander L L Johnston was an RAF officer.

Equipped with Vickers Wellington IV bombers, 458 Squadron participated in its first operation on October 20 when ten aircraft joined in attacks against Emden, Antwerp and Rotterdam. While these raids were successful, the targets proved to be well defended and Sergeant P Hamilton's Wellington failed to return while another three bombers were damaged.

Night attacks against German and French targets continued, at increasing intensity as the unit gained in experience and confidence. Early

A bomber crew and ground staff of 458 Squadron in the United Kingdom. Wellington "L For Leather" is the backdrop. (RAAF Museum)

70 Squadrons based at Abu Sueir or to 104, 108 and 148 Squadron at Kabrit. All of these were RAF squadrons.

The detached aircrews commenced night bomber operations with their RAF squadrons and while so engaged many of them almost completed an entire operational tour of duty. During one mission in July, Pilot Officer C Hare force landed his Wellington 700km behind enemy lines. Four crew members decided to evade capture and head for friendly territory. Travelling in the cool of the evenings over a period of 22 days they eventually were found by Allied troops in what was one of the most remarkable escapes of World War II.

Finally, in September 1942, 458 Squadron was reunited at El Shallufa and commenced shipping searches, anti-submarine patrols, convoy escorts and mine laying missions. This type of work resulted in a considerable number of detachments being made, the largest being to Malta. Enemy shipping was regularly located and attacked by the Wellingtons – a number being damaged or sunk.

Late in March 1943, 458 Squadron relocated to LG 91 and from here operations continued at a high pitch and Wellington losses, while not excessive, rose steadily. Between June and August, 174 sorties were flown and seven ships sunk, including one of 7140 tonnes, destroyed by Flight Sergeant Bishop's crew and another of 11,220 tonnes sunk by a Wellington captained by Pilot Officer Quinlan. In July alone, one merchant ship was sunk and an Italian cruiser and destroyer damaged. On the debit side, however, no less than four Wellingtons were lost.

In October 458 Squadron moved to Bone in Algeria where it conducted anti-submarine patrols and convoy escort missions. These were usually uneventful, however, several U Boats were sighted and at least two were damaged. After again moving, this time to Alghero (Sardinia) in May/June 1944, the unit gained its first submarine kill on the night of May 14/15 when a Wellington captained by Squadron Leader R Knights sighted and attacked U-731 shortly after it submerged. Oil came to the surface and it was later confirmed that the submarine had indeed sunk.

458 Squadron moved to Foggia in September 1944 with a detachment almost immediately deploying to Falconara. By late January 1945 the unit had again relocated, this time to Gibraltar where it remained until the cessation of hostilities.

To celebrate the Allied victory, 458 Squadron

targets included Mannheim, Aachen, Cologne, Dusseldorf and Bremen. All of these industrial cities were well defended and it was a regular occurrence for the ground staff to spend long days repairing flak and fighter damaged bombers. Just as often, however, crews simply disappeared – victims of an extraordinarily effective defence system which encompassed virtually all of occupied Europe. In the early days the Wellingtons also undertook a heavy mine laying programme – sowing mines along enemy occupied coasts.

In November 1941 a flight of this by now seasoned unit became the nucleus of the newly formed 460 Squadron while, in early 1942, 458 Squadron faced further upheaval when it was withdrawn from Bomber Command and reassigned to the Middle East.

The first Wellingtons departed on February 22 1942 in a deployment which could at best be described as chaotic. The bombers were to stage through Malta but once there they were commandeered and their crews stranded for weeks awaiting replacement aircraft. To make matters worse, the Commanding Officer and his crew were shot down by enemy fighters near Malta and only the rear gunner, who destroyed one of the attackers, survived as a POW.

When the ground staff finally arrived in the Middle East in May they found that the aircrew had been allocated to other units and they themselves were split up and put to work at various airfields servicing amongst other aircraft, USAAF Liberators and RAF Wellingtons. Meanwhile, replacement aircrews posted to 458 Squadron had gathered in Britain (at Morton-in-the-March) but, on arriving in the Middle East, were attached with their Wellingtons to 37 and ground personnel participated in a parade while the Wellingtons flew in formation overhead. This non operational flight was especially frightening to many of the aircrews who were out of practice in formation flying. 458 Squadron disbanded at Gibraltar on June 9 1945.

Formed: *July 8 1941*
Squadron Code: *MD*
Aircraft: *Wellington*
Locations: *Williamtown: July 1941 – August*
Holme-on-Spalding-Moor: Aug 1941 – Mar 1942

Kasfareet: May 1942
Fayid: May – June 1942
St Jean: July – August 1942
Aqir: August – September 1942
Shallufa: September 1942 – May 1943
LG 91: March – June 1943
Protville l: June – October 1943
Bone: October 1943 – June 1944
Alghero: June – September 1944
Foggia: September 1944
Gibraltar: January – June 1945
Disbanded: *June 9 1945*

---

459 SQUADRON

Under the command of Squadron Leader P W Howson, 459 Squadron formed in Egypt at Burg-El-Arab on February 10 1942.

The unit, which was to have been equipped with Lockheed Hudson IIIs to operate in the maritime reconnaissance role, initially found itself with just two of these aircraft and four borrowed Bristol Blenheim IVs.

459 Squadron's first operation was conducted on February 14 when the two Hudsons, one of which made an unsuccessful attack on a submarine, were despatched on reconnaissance missions. In its first few weeks 459 Squadron flew convoy escorts and anti-submarine patrols with its small fleet of aircraft. By May, however, the unit had received its full allocation of Hudsons and was fully operational.

In June the unit began an intensive campaign against German shipping, particularly F Boats – heavily armed tank landing craft. On July 28 four Hudsons made their first attack on two F Boats. One of the enemy vessels was so badly damaged that it was beached, however, the cost was high – one Hudson shot down and another damaged. In July alone, 12 of these craft were sunk but at heavy cost – no less than six Hudsons. Further heavy losses were inflicted on the enemy's shipping in August and the Hudson crew's efforts were rewarded when the Germans withdrew the surviving F Boats from their vitally important cargo carrying tasks. Moving to a new airfield at Idku, patrols and strikes against enemy shipping continued over the succeeding months.

To cover 459 Squadron's vast area of responsibility, detachments at Aden, St Jean and later Nicosia and Berka, were maintained at various times. While many missions emanating from these locations were of a routine nature, some major successes were attained. In September 1942 Flying Officer Beaton's crew left a destroyer on fire and sinking, while, on June 16 1943, a Hudson captained by Flight Sergeant D T Barnard made a wave top height attack on a submarine. The Hudson's depth charges destroyed the submarine (U-97) but the blast effect so badly damaged the aircraft that it only just managed to reach land where a crash landing was effected.

In September 1943, 459 Squadron temporally changed roles to that of a bomber unit and conducted day and night strikes against targets on Greece and Crete. In February 1944 the unit re-equipped with Lockheed Ventura Vs after which it relocated to Gambut and later Ramat David. The new aircraft were not successful, and after operating against enemy shipping around Rhodes and Leros Islands, were replaced by Baltimore IVs in July.

Anti-shipping patrols and bomber attacks around the Greek Islands continued until February 1945 when 459 Squadron moved to England to re-equip with Wellingtons. This plan did not eventuate and, still non operational, 459 Squadron disbanded at Chivenor on April 10 1945.

Formed: *February 10 1942*
Squadron Code: *BP*
Aircraft: *Blenheim, Hudson, Ventura, Baltimore*
Locations: *Burg-el-Arab: February – May 1942*
LG 40: May – July 1942
Gianaclis: July – September 1942
LG 208: September – November 1942
Gianaclis: November – December 1942
Gambut 3: December 1942 – April 1944
Ramat David: April – May 1944
St Jean: May – August 1944
Berka: August 1944 – February 1945
Chivenor: March – April 1945
Disbanded: *April 10 1945*

---

460 SQUADRON

Under the Command of Wing Commander A L Hubbard, 460 Squadron formed at Molesworth on November 15 1941. Equipped with twin-engined Wellington IVs and trained in a night bomber role, the unit moved to Breighton in January 1942 and flew its first mission on March 12 when five Wellingtons participated in an attack on Emden.

460 Squadron's first operational loss occurred on March 13 when a Wellington, captained by Sergeant J F D Cooney, failed to return from a raid against Dunkirk. In this period a number of leaflet dropping missions over Paris and other French cities were made, however, within a short time, attacks on heavily defended German targets became the norm rather than the exception.

460 Squadron gained its first aerial victory early in its existence – on the night of April 25/26 when Sergeant S Levitus's Wellington was attacked by three Bf 110 night fighters. The bomber's tail gunner shot down one of the attackers, thus enabling his bomber to escape almost certain destruction. In May the unit despatched 18 Wellingtons to Cologne in Bomber Command's first 1000 bomber raid of the war. This devastating raid left much of the city destroyed.

However, by 1942 German fighter and anti-aircraft defences were well organised and active. Between June and August of 1942, 460 Squadron lost 20 Wellingtons with most of their crews. Many other aircraft were damaged and their repair was a difficult task for the ground staff who had to keep the maximum number of bombers available for operations. Despite its losses, 460 Squadron maintained one of the highest – if not the highest – serviceability rates in Bomber Command.

460 Squadron's last Wellington operation was flown on the night of August 27/28 1942, when it participated in an attack on Kassel. Two Wellingtons were lost and another, captained by Sergeant P S Isaacson, returned severely damaged by night fighters. Just over two weeks later, the unit began converting to Handley Page Halifax B.II four engined heavy bombers. Despite improved performance, the Halifaxes were not popular with their crews and within a few months were replaced by Avro Lancaster B.I and B.IIIs.

Lancaster operations commenced on November 22 when nine aircraft – one of which failed to return – were despatched to Stuttgart. Many of 460 Squadron's attacks were against the Ruhr Valley and Berlin – especially heavily defended targets. Accordingly, losses were heavy. A number of attacks against lightly defended Italian and French targets were also conducted.

On May 14 1943, 460 Squadron moved to Binbrook where operations recommenced two days later. With some 30 Lancasters divided into three flights the unit continued to pound German targets in particular. On June 11 no less than 27 Lancasters – a squadron record – were despatched against Dusseldorf. Later, in August, 460 Squadron became the first Bomber Command squadron fly 1000 sorties in Lancasters.

460 Squadron's greatest single loss of the war occurred on the night of December 2 1943 when, out of 24 Lancasters despatched against Berlin, five failed to return, one crashed on landing and a further three were damaged. Despite the loss of these aircraft and personnel, the following night, 16 Lancasters raided Lipzip. On December 16 a further four Lancasters were lost in crashes on returning from a mission when Binbrook was covered in fog.

On the night of January 2/3 1944, Berlin was successfully attacked despite a disaster at Binbrook that jeopardised 460 Squadron's participation in the raid. On its takeoff run, a fully laden Lancaster crashed into Binbrook village killing all its crew. Despite the demoralising effect of taking off in the glow of fires caused by the crash, the remaining bombers got away and successfully attacked their target.

By early 1944 the Lancasters were being used to hammer coastal fortifications and other French targets in preparation for Operation "Overlord", the Allied invasion of Europe. V-1 flying bomb launching sites and assembly areas were also attacked.

460 Squadron broke its own Bomber Command record on

An Australian Lancaster returns after a night mission.

May 24 1944 when it dropped 136 tonnes of bombs on coastal defences at Le Clipton. Next month the first daylight raid in the Squadron's history was conducted when 20 Lancasters joined in an attack on Le Havre.

In August 460 Squadron again broke a Bomber Command record – which, incidentally, it already held – by delivering 1904 tonnes of bombs during the month. The remainder of 1944 and into 1945 saw a variety of targets attacked by the Australian bombers, generally at a lower loss rate as the German defences progressively became disorganised.

Flying Officer E C Owen's crew had a lucky escape on the night of November 9/10 whilst attacking Wanne-Eickel. Badly damaged during the raid, Flying Officer Owen was seriously wounded and incapacitated. The Lancaster's bomb aimer, Pilot Officer L W Woods, took control and not only flew the damaged aircraft to England, but made a safe landing at the end of his long flight!

On the night of February 23 1945, 24 Lancasters participated in a successful attack against Pforzheim. One bomber, captained by Flying Officer J M Cox was attacked by two jet fighters which in a remarkable display of marksmanship were both shot down by the Lancasters gunners, Flight Sergeant G W D Crosby and Flying Officer A M Curren. February was an especially busy month for 460 Squadrons ground staff as, in one 13 night period, the unit participated in 14 major attacks.

460 Squadron flew its final mission of the war on April 25 1945 when 20 Lancasters were despatched against the Berchtesgaden – Hitler's mountain retreat. The target was totally destroyed. One Lancaster, captained by Flying Officer H G Payne, was shot down by anti-aircraft fire, however, its crew escaped.

With Germany's surrender the following month, 460 Squadron's aircraft were used to repatriate Allied POWs from Germany and later, to drop food supplies to starving civilians in Holland. Selected as one of the Bomber Command squadrons to operate in the Pacific against the Japanese, 460 Squadron moved to East Kirkby late in July and commenced training for its new role. Japan's unexpected capitulation the following month put paid to these plans, however, and 460 Squadron disbanded on October 2 1945.

During its 6264 operational sorties 460 Squadron gained the reputation of being Bomber Command's foremost squadron. This reputation was achieved at an enormous cost however. 188 aircraft and no less than 978 Commonwealth airmen lost their lives while serving with the unit. A tangible reminder of this sacrifice, and indeed the sacrifice of all Australians who served in Bomber Command, survives in the form of "G for George" – one of the unit's Lancasters – which may be seen and admired at the Australian War Memorial in Canberra.

Formed: *November 15 1941*
Squadron Code: *UV, AR*
Aircraft: *Wellington, Halifax, Lancaster*
Locations: *Molesworth: Nov 1941 – Jan 1942*
Breighton: January 1942 – May 1943
Binbrook: May 1943 – July 1945
East Kirkby: July – October 1945
Disbanded: *October 2 1945*

---

461 SQUADRON

461 Squadron formed at Mount Batten on April 25 1942. Allocated an anti-submarine role, the unit was to have been equipped with Catalinas, however, Sunderland Mk.IIA flying boats were substituted instead.

Patrols commenced on July 1, however, it was not until September 1942, after 461 Squadron moved to Hamsworthy, that contact was made with enemy submarines. In this month, eight attacks were made and several submarines damaged. During this period one Sunderland, the first of many, was shot down by German fighters over the Bay of Biscay.

Over the succeeding months, more attacks were carried out, however, it was not until May 2 1943 that 461 Squadron achieved its first confirmed U Boat kill when Flight Lieutenant J C Smith's crew located a surfaced submarine. As the Sunderland dived to attack, its nose gunner strafed the German gun crew, allowing his captain to make an accurate attack which resulted in the destruction of U-332.

On May 29 Pilot Officer Singleton's crew landed on the open sea to rescue survivors from two aircraft. Due to the heavy swell, and the weight of the additional passengers, the Sunderland could not immediately takeoff, and the survivors were later transferred to a destroyer which had arrived at the scene. The flying boat then attempted a hazardous takeoff but in the process sustained damage to an engine and more seriously, had a one metre long by eight centi-

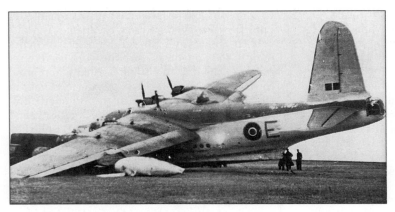

One of the most famous photos to come out of World War II – Flying Officer G Singleton's successful landing on Angle airfield after his flying boat had been badly holed whilst taking off from the open sea.

successful and well designed that they were later adopted throughout the RAF.

On July 30 Flight Lieutenant D Marrow's Sunderland was in a mixed force of aircraft, and later surface vessels, which engaged three U Boats. All three submarines were sunk; curiously enough the 461 flying boat, coded U/461, sank the submarine U-461. On the way home from their successful engagement the Sunderland crew located a second submarine, but this time their flying boat was badly damaged by return fire before being able to release its remaining depth charges. The crippled flying boat made a forced landing at the Scilly Islands without injury to the crew.

461 Squadron's last major success for 1943 occurred on August 2 when the U-106 was sent to the bottom by Flight Lieutenant I A F Clark's crew in conjunction with a Sunderland of 228 Squadron RAF. For the remainder of the year, routine patrols, some of which were bitterly contested by the Luftwaffe, were conducted and the next U Boat kill did not occur until January 28 1944 when the U-571 was destroyed.

1944 also saw 461 Squadron operating in a new role – that of night strike using radar equipment and Leigh lights. Despite this task, anti-submarine patrols remained the unit's 'bread and butter' and resulted in the destruction of U-385 on August 11, while next day 461 Squadron's last submarine victim, U-270, was sunk.

For the remainder of the war, patrols were conducted, however, even after the Sunderlands had been fitted with sonobuoy submarine detection equipment in April 1945, no further sinkings were made prior to Germany's surrender. 461 Squadron disbanded at Pembroke Dock on June 20 1945.

metre high gash torn in its hull. Due to the serious hull damage, the flying boat was unable to realight on the sea, Pilot Officer Singleton elected to land on Angle airfield. After jettisoning fuel, other flammable and surplus equipment, the pilot made a smooth landing cutting a deep furrow through the grass strip with the Sunderlands keel. The crew escaped injury and the only additional damage caused was to the flying boat's port wing float.

Three days later, Sunderland N/461 captained by Flight Lieutenant C B Walker was intercepted over the Bay of Biscay by eight Junkers Ju 88 long range fighters. In the epic battle which followed, three fighters were destroyed, another two probably destroyed and the remaining three damaged and forced to abandon the combat. The bullet riddled flying boat, with five wounded crewmen on board (one of whom later died) just reached the Cornish Coast and force landed in the shallows where it promptly sank.

This engagement was one of the most extraordinary in the annals of Coastal Command and moved the RAF's Chief of Air Staff, Air Chief Marshal Slessor, to write to 461 Squadron's Commanding Officer in the following terms:

"I should like Flight Lieutenant Walker and the surviving members of his crew to be told of the admiration and pride which I felt on reading the details of this epic battle which will go down in history as one of the finest instances in the war of the triumph of coolness, skill and determination against overwhelming odds ..."

An important factor in the survival of this crew was the inventiveness of 461 and 10 Squadron ground staff who modified their Sunderlands with twin gun nose turrets and 'galley' mounted machine guns. These modifications proved so

Formed: April 25 1942
Squadron Code: UT
Aircraft: Sunderland
Locations: Mount Batten: April – August 1942
Hamsworthy: August 1942 – April 1943
Pembroke Dock: April 1943 – June 1945
Disbanded: June 20 1945

A unit destined to have one of the most chequered histories in the RAAF, 462 Squadron formed at Fayid, Egypt on September 7 1942 from detachments of 10 and 76 Squadrons RAF. The irregular manner in which this "RAAF" squadron came into existence ensured that very few Australians served with the unit – initially only one aircrew member and just a handful of ground staff were RAAF members.

Equipped with Halifax bombers, the unit, commanded by Wing Commander D O Young, flew its first operation – an attack against Tobruk – on the night of September 8/9. Operations continued with some success up until December when the unit became non operational due to a lack of replacement aircrews. In the interim, 462 Squadron had operated from several airfields and was now based at Solluch. As the only Halifax squadron in the Middle East it became necessary for the unit to train its own replacement aircrews.

Parallel to 462 Squadron's operations during 1942, RAAF Overseas Headquarters had been making strong representations to the RAF in an attempt to obtain more Australian personnel posted to the unit. These attempts met with little success and Wing Commander Young was later forced to scour the Middle East in search of surplus ground staff of any trade which he sometimes had to misemploy to give the unit a more Australian character.

These efforts had mixed results as the newly arrived Australians invariably knew nothing about Halifax maintenance, the result being that serviceability fell for a period. The majority of aircrew, who were mostly RAF, resented having their deeds recorded in the annals of the RAAF rather than their own service and relations between the few Australians and their Commonwealth comrades were, at times, strained. As late as August 1943, only 160 of 462 Squadron's 600 personnel were RAAF members.

Operational again by the beginning of January 1943, the unit moved to Gardabia Mam in Cyranaica the following month where strikes against targets in Sicily were undertaken. Later Rhodes, Piraeus and Salamis Islands also came in for their share of night attacks. Further moves occurred during 1943 and by early 1944, 462 Squadron was split – part of it operating from Foggia, Italy, with the remainder in Libya.

On the night of May 6/7 1943 a Halifax captained by Warrant Officer N Vertican was forced down in the sea 130km west of Tripoli. The pilot made an excellent ditching in which all crew members had time to escape into their dinghy. After drifting at sea for ten and a half days with very little to drink or eat, the crew (which included just one Australian, Sergeant C D Curnow) reached the shore where they were rescued by Arabs and handed over to Allied troops. During a similar incident in July, another crew drifted for nine days before being picked up by a passing destroyer.

At this point in time, RAAF Overseas Headquarters became so exasperated with the RAF's duplicity in posting so few RAAF members into 462 Squadron that it requested that the unit be disbanded and reformed in the United Kingdom. This request, unlike those previously made, was acted on with alacrity and, on March 3 1944, 462 Squadron RAAF was redesignated 614 Squadron RAF.

Not dormant for long, 462 Squadron reformed at Driffield in the UK on August 12 1944. Again equipped with Halifaxes, this time the unit was given an Australian Commanding Officer and a much greater proportion of Australian air and ground crews. Operational within days, 462 Squadron participated in day and night attacks against German industrial cities while at the same time supporting the Allied ground forces fighting their way across France.

An early Merlin powered Halifax with nose turret. (RAAF Museum)

After just five months operating as a main force bomber unit, 462 Squadron moved to Foulsham in December where it joined 100 (Bomber Support) Group. This Group specialised in disrupting the German air defence system utilising radio countermeasures and other means to achieve its goals.

462 Squadron's aircraft were modified to carry special radar jamming equipment designed to interfere with both the night fighters and ground based radar. Until all the aircraft had been modified the crews dropped aluminium strips (known as "window") from their aircraft. Window shaped to the right frequency produced a radar image of an aircraft and if enough of it was dropped, an entire attack could be simulated by a handful of aircraft. To further add to this effect, 462 Squadron began to carry small loads of incendiaries, target markers (as used by the Pathfinders) and bombs which were dropped to further confuse the enemy.

The role of units such as 462 Squadron was absolutely crucial to the success of the bomber offensive. The Germans were frequently bluffed into believing that full scale attacks were being made on nights when, in fact, no operations were planned. On other occasions, the German night fighters were lured away from the real bomber stream onto an area which consisted of nothing but a few aircraft dropping thousands of bundles of window.

Protected by their own countermeasures, the Australian Halifax squadron suffered very light losses during 1945 despite its intensive operations and the fact that on some nights it was part of very small formations operating over occupied territory. In February 1945 alone, the unit, operating on 15 nights and one day, despatched 155 aircraft. On one mission when five Halifaxes were over the Ruhr the only real difficulty encountered was an attack by one night fighter on Pilot Officer A D J Ball's aircraft. Fortunately the bomber's gunners had the better of the encounter and badly damaged the fighter which abandoned its attack.

462 Squadron's last wartime operation was flown on the night of May 2/3 when 10 Halifaxes raided Flensburg to cover a real attack against Kiel. All aircraft returned safely. After the German surrender, the unit continued flying for a time, mainly conducting ferrying duties and training flights. A number of armed patrols were also flown by day over Germany. 462 Squadron disbanded at Foulsham on September 24 1945.

Formed: *September 7 1942*
Squadron Code: *R, Z5*
Aircraft: *Halifax*
Locations: *Fayid: September – November 1942*
LG 237: November 1942
LG 167: November – December 1942
LG 237: December 1942 – January 1943
Solluch: January – February 1943
Gardabia Mam: February – May 1943
Hose Raui: May – October 1943
Terna: October 1943 – January 1944
El Adam: January – March 1944
Disbanded:
Driffield: August – December 1944
Foulsham: December 1944 – September 1945
Disbanded: *September 24 1945*

---

463 SQUADRON

Under the command of Wing Commander R Kingsford-Smith (a nephew of Australia's most famous aviator – Sir Charles Kingsford-Smith), 463 Squadron formed at Waddington on November 25 1943. For its nucleus, the new Bomber Command squadron received an allocation of Avro Lancaster B.I and B.IIIs and personnel from 467 Squadron's "C Flight".

Operations began next night when six aircraft participated in an attack on Berlin. Strong defences were responsible for the loss of one Lancaster, however, another crew claimed the unit's first victory when they shot down a Ju 88 night fighter. As 463 Squadron received more aircraft and personnel the weight of its attacks steadily increased. Operations against German targets, however, were made against strong fighter, anti-aircraft and search light defences and were only accomplished with considerable loss.

In another raid against Berlin on the night of January 2/3 1944, for example, two Lancasters were lost, one crew being killed while five of the seven members of the second crew survived as POWs. It should be noted, however, that most airmen perished in their aircraft while some of those who did manage to bale out were murdered by enraged German civilians when they landed in the target area. Another bad night occurred on January 29/30 when three Lancasters were lost – again over Berlin.

On the night of May 3/4, 463 Squadron participated in a very successful attack against a

German tank division in France. The bombing was so accurate that almost all the division's vehicles and support equipment was destroyed and many personnel killed or wounded. Just a week later, on May 10/11, the unit suffered its greatest single loss of the war when it took part in an attack against the German airfield at Lille. Facing bitter fighter opposition, no less than four Lancasters were shot down while several others were damaged. One crippled aircraft, captained by Squadron Leader W Brill, returned with one engine almost completely shot away.

In the lead up to the Allied invasion of Europe, 463 Squadron attacked targets throughout France and on the night of June 5/6, as the landing was about to commence, bombed German coastal batteries on the landing beaches. After the landings, further attacks in support of the ground forces were mounted. June also saw 463 Squadron participate in its first daylight mission when Beauvoir was attacked and during the next 10 months such attacks became more common as the German fighter defences progressively became disorganised and less efficient and the quality of the escorts increased.

463 Squadron also regularly supported the activities of the RAF Film unit whose task was to record special operations. Aircraft designated for this special role were modified with cameras in their front turret and another near the crew access door. These Lancasters participated in regular attacks and also during special operations such as the D Day landings.

While carrying Film unit personnel one Lancaster, captained by Flight Lieutenant B A Buckham, was given the unusual task of recording a series of attacks against the battleship *Tirpitz* in Tromso Fiord, Norway. The third raid on November 12 severely damaged the ship which, after the bombing force had departed, slowly capsized. The 463 Squadron crew, who were just leaving the area, returned and captured the *Tirpitz'* last moments. The total mission time was 14½ hours and was one of the longest – if not the longest – undertaken by a Lancaster.

A most unusual incident occurred during an attack against Giessen on the night of December 6/7. A 463 Squadron crew observed a Bf 110 about to attack another Lancaster. The captain, Flying Officer Bennett, ordered his gunners to open fire and the German fighter crashed in flames. It was indeed unusual for the vulnerable Lancasters to become the aggressor and furthermore, to achieve such results with their light defensive armament.

1945 saw attacks continue against declining enemy opposition. Apart from strikes on German industrial centres and tactical targets in support of the Allied armies, the Australian Lancasters operated against some unusual targets which included the well defended Dortmund-Ems Canal – a vital German transport system – which was severely damaged on a number of occasions.

During a daylight attack against Hamburg on April 9, the unit had its first brush with German jet fighters and this was followed less than two weeks later by its last wartime operation. On this occasion 14 aircraft attacked an oil refinery at Tonsberg on the night of April 25/26. One bomber, captained by Flying Officer A Cox crash landed in Swedish territory and its crew interned until the end of hostilities – a matter of days. Even before the war had ended, 463 Squadron flew liberated POWs to England.

After moving to Skellingthorpe early in July, 463 Squadron disbanded on October 29 1945. Proportionally, 463 Squadron sustained the highest loss rate of any Australian squadron in the European theatre of war.

Formed: *November 25 1943*
Squadron Code: *PO*
Aircraft: *Lancaster*
Locations: *Waddington: Nov 1943 – July 1945*
Skellingthorpe: July 1945 – October 1945
Disbanded: *October 29 1945*

Night takeoff for a 463 Squadron Lancaster.

464 SQUADRON

Equipped with Lockheed Ventura I light bombers, 464 Squadron formed at Feltwell in the United Kingdom on September 1 1942.

The new unit, commanded by Wing Commander W C Young, participated in its first operation on December 6 when 14 Venturas attacked Eindhofen. Although three bombers were lost to anti-aircraft fire and another was badly damaged, bombing was very accurate and considerable damage to the target resulted.

464 Squadron moved to Methwold on April 2 1943 where it joined the only other two Ventura squadrons in Bomber Command (21 Squadron RAF and 487 Squadron RNZAF) and continued its operations, with moderate success, against French targets. These attacks usually involved high or medium level formation bombing and, as the bombers usually had strong fighter escort, those losses sustained were solely the result of anti-aircraft fire.

Unfortunately, an incident involving 487 Squadron sealed the fate of the Ventura's service over Europe. On May 3, 11 New Zealand aircraft scheduled to attack a target in occupied Europe failed to make contact with their fighter escort but pressed on to their target area anyway. The German fighters, only having to contend with the defensive fire of the lightly armed, slow and unmanoeuvrable bombers shot ten of them down. After this disastrous mission, plans were made to replace the Venturas but operations continued until July 10 when 464 Squadron's last Ventura mission – an attack against St Omer airfield – was flown. Eleven days later, 464 Squadron moved to Skulthorpe and converted to Mosquito VIs.

464 Squadron flew its first Mosquito mission when 12 aircraft bombed Mur-De-Bretagne power station on October 3 1943. Despite heavy anti-aircraft fire which damaged several Mosquitos, the station was successfully hit and seriously damaged.

Intruder and attack operations continued by day and night, not only from Skulthorpe, but also from Ford (from where several Ju 88s were shot down over the Bay of Biscay) and Bradwell Bay. Later, in December 1943, 464 Squadron again moved, this time to Hunsdon.

From 1944 intensive operations against V-1 launching sites and storage facilities commenced. These targets were well defended and, due to their small size, difficult to hit. Consequently, special tactics had to be evolved and even so losses, at times, were heavy while attacking these so called "vengeance" weapon sites.

During one attack against a V-1 launch ramp, Flying Officer G Oxdale's Mosquito was hit by anti-aircraft fire and apart from other damage, it had over two metres of its starboard wing blown off. It was only with great difficulty that this pilot was able to recross the English Channel and make a forced landing at an emergency landing ground. In the process, the crippled Mosquito was so badly damaged that it was later written off, however, the crew escaped injury.

Having gained a reputation for extremely accurate attacks, 464 Squadron – along with 21 and 487 Squadrons which had also converted to Mosquitos – was selected to attack the Amiens Prison where hundreds of French Resistance fighters were about to be executed. Code named Operation Jericho, 464 Squadron despatched six aircraft to attack the guards barracks while the other squadrons breached the prison walls. The raid took place on February 18 1944 and, while some prisoners died in the bombing, others made good their escape. Two RAAF aircraft failed to return and only one of the four airmen involved – Squadron Leader I McRitchie – survived as a POW.

In support of the Allied invasion of Europe, 464 Squadron operated at maximum strength on the night of June 5/6 1944, despatching no less than 20 Mosquitos against road traffic in the vicinity of the landing sites. After the bridgehead had been established intensive night operations, chiefly against communications targets, continued.

464 Squadron Venturas at Methwold. Note the kiwi and emu nose art. (RAAF Museum)

On the night of July 14/15, four Mosquitos along with others from 21 and 487 Squadrons, attacked and destroyed Gestapo billets at Bonneiul-Matours. This was the first of several precision attacks against the Gestapo in which these Squadrons participated. On October 31 an attack against Gestapo Headquarters at Aahus, Denmark, killed some 200 enemy personnel and destroyed valuable records. 464 Squadron moved to Rosieres, France, in February 1945 where it continued damaging attacks on the enemy, both by day and night.

March 21 saw the next of 464 Squadron's attack against the Gestapo when six Mosquitos, along with a similar force from 21 and 487 Squadrons, attacked another headquarters in Denmark – this time at Copenhagen. Despite the loss of two RAAF aircraft, the raid was highly successful, the building being gutted by fire.

Thirteen aircraft were despatched on attack missions on May 2 1945. These proved to be the last operational sorties flown before Germany's collapse. After the surrender 464 Squadron conducted a number of victory flypasts and on one occasion escorted British Prime Minister Winston Churchill's aircraft to Bordeaux. Another task undertaken was to fly General Jodl, the German officer who signed the surrender agreement, to Berlin.

464 Squadron disbanded at Melsbroek, Belgium on September 27 1945.

Formed: September 1 1942
Squadron Code: SB
Aircraft: Ventura, Mosquito
Locations: Feltwell: Sept 1942 – April 1943
Methwold: April 1943 – July 1943
Skulthorpe: July 1943 – December 1943
Hunsdon: December 1943 – February 1945
Rosieres: February 1945 – April 1945
Melsbroek: April 1945 – September 1945
Disbanded: September 27 1945

---

466 SQUADRON

Under the command of Wing Commander R E Bailey and equipped with Vickers Wellington Mk.III bombers, 466 Squadron formed at Driffield, England on October 10 1942.

Moving to Leconfield towards the end of November, the unit flew its fist operation – a mine laying mission – on January 13 1943. This operation was the first in a long series of minelaying operations which were carried out parallel to an increasing night bombing commitment. The next evening, during another minelaying operation the unit sustained its first combat loss when Sergeant R V Babbington's crew failed to return.

On January 15, four crews joined a force which attacked Lorient docks – the unit's first bombing attack. Two weeks later, on January 30, 466 Squadron's first attack against a German target was made when eight Wellingtons were despatched against Emden. This attack was most unusual in that it was made in daylight, the vulnerable bombers being protected by thick cloud for most of their time over enemy territory. Notwithstanding this natural protection, two aircraft were lost.

On the night of February 19/20, Sergeant R Rosser's tail gunner, Sergeant H Wilcock, claimed 466 Squadron's first aerial victory when he downed a Bf 110. Return fire from the German aircraft wounded two crew members and set the bomber on fire. Thinking that all was lost, the bomb aimer baled out, however, the fire was bought under control and the aircraft successfully returned to England.

Sadly, aerial victories such as these were not commonplace and, as the German night fighter and anti-aircraft defence system continued to develop, an increasing number of Wellingtons were lost – especially over Germany. Other hazards regularly encountered included bad weather conditions, the risk of mid air collisions and being hit over the target by the bombs dropped from other aircraft.

On the night of April 14, a Wellington captained by Sergeant E F Hicks was attacked by a night fighter and extensively damaged. The tail gunner, Sergeant R F Field, was killed and three other crew members wounded. Despite the serious damage to their aircraft and their wounds, the crew elected to press on and bomb their target – Stuttgart. After this had been accomplished, the long return flight to England was made where an emergency landing at Ford was effected.

This crew's tenacity was rewarded with the award of the Conspicuous Gallantry Medal to Sergeant Hicks for his determination and three of this pilot's four surviving crewmen were also decorated – one each with the DSO, the DFC and the DFM.

It was most unusual for so many personnel from one crew to be decorated and their awards represented the first made to members of 466 Squadron.

Wellington operations continued at an intensive rate until late August when the unit converted to Handley Page Halifax Mk.IIs. Once fully re-equipped and retrained, 466 Squadron was operational again by late November and flew its first Halifax mission on December 1 when 12 aircraft laid mines off Terschelling Island.

466 Squadron Halifaxes were fitted with H2S, a radar navigational and bombing aid which considerably enhanced the aircraft's ability to locate and attack their targets. 466 Squadron was the first non-Pathfinder Force squadron to receive this equipment.

The unit's first attack on a German target with Halifaxes occurred on the night of December 20/21 1943 when 16 aircraft joined with 634 other bombers in an unsuccessful attack on Frankfurt. Two crews were lost and a third returned to England in a badly damaged aircraft, the result of a determined night fighter attack.

From May 1944, some of 466 Squadron's bombing effort was directed against "invasion" targets in France – coastal batteries, marshalling yards and military bases. These targets were a welcome relief for the aircrews as, lacking the same degree of defence encountered over Germany, losses tended to be lower.

In early June, 466 Squadron moved back to Driffield, where intensive operations against German targets and in support of the imminent Allied landings in France continued. On the night of June 5/6, while returning across the English Channel, after attacking coastal batteries, crews witnessed the spectacle of the invasion armada as it approached the French coast.

During a daylight attack on June 22, 466 Squadron struck a V-1 site at Siracourt. Anti-aircraft defences were especially active and in the heavy barrage over the target one Halifax was shot down and a further six damaged. With the war progressing to its inevitable conclusion, further daylight missions were mounted as it became apparent that the German Air Force was no longer able to operate in strength against the Allied air forces.

On the night of August 12/13, 466 Squadron participated in an attack on Brunswick in which only scattered bombing resulted due to poor weather. Enemy night fighters were active over the target and many bombers were lost. However, 466 Squadron suffered no losses despite attacks made on several aircraft, and the bombers gunners claimed three night fighters destroyed.

During an attack on Dusseldorf on the night of November 2/3, Flying Officer A Kopp's crew had a remarkable escape then their bomber was hit by a load of incendiaries from another aircraft. The Halifax was badly damaged but reached England, where, next day, an unignited incendiary was discovered in one of the bomber's fuel tanks.

One of the most remarkable escapes of the war occurred on the night of November 4/5 when Pilot Officer J Herman's Halifax was hit by anti-aircraft fire. After ordering his crew to bale out, the captain was just about to clip on his parachute when the crippled aircraft exploded. Flying Officer Herman was blown out of the aircraft without a parachute and, while plummeting to the ground in total darkness, bumped into something which he instinctively grabbed. The "something" turned out to be his mid upper gunner's legs. Remarkably, both airmen made a safe descent on the one parachute and survived the rest of the war as prisoners.

Despite the onset of the worst winter Europe had seen in 50 years, operations continued at a hectic pace into 1945. After returning from an attack on Kamen on the night of March 3, German night intruders operating over the United Kingdom caught some of 466 Squadron's aircraft as they neared Driffield and two Halifaxes were shot down.

These losses were deeply felt as, after completing a long and dangerous mission over Europe, the returning crews had always felt themselves safe after crossing the English Channel.

466 Squadron's Halifax "HD-J". (RAAF Museum)

On Anzac Day 1945, 466 Squadron flew its final operational mission – an attack against coastal defences on Wangerooge Island. After the German surrender, 466 Squadron became part of Transport Command and for a period jettisoned surplus bombs into the sea. On June 20 1945 466 Squadron was renumbered 10 Squadron but plans for the unit to re-equip with Lancasters and Yorks did not eventuate and the unit disbanded soon afterwards.

Formed: *October 10 1942*
Squadron Code: *HD*
Aircraft: *Wellington, Halifax, Liberator*
Locations: *Driffield: Oct 1942 – Nov 1942*
Leconsfield: November 1942 – June 1944
Driffield: June 1944 – June 1945
Bassingbourn: June – June 1945
Disbanded: *June 20 1945*

467 SQUADRON

Commanded by Wing Commander C L Gomm, 467 Squadron formed at Scampton on November 7 1942. Equipped with Avro Lancaster B.I and B.IIIs, the unit operated over Europe for the duration of hostilities.

While training 467 Squadron moved to Bottesford at the end of November and conducted its first operation on the night of January 2/3 1943 when five Lancasters participated in a mine laying mission off the French coast. The first bombing attack to Germany – against Essen – was flown three nights later.

Attacks at increasing strength were made against German, and to a lesser extent, French and Italian targets. Despite strong enemy opposition, 467 Squadron did not sustain its first operational loss until May 25/26 1943 when two out of 18 Lancasters, despatched in a damaging attack against Dusseldorf, failed to return.

Despite losses which rapidly mounted after this date, 467 Squadron gained a reputation for pressing home its attacks, which included a considerable number against heavily defended Berlin. During an attack against Mannheim on August 9/10 intense night fighter opposition was encountered with four Lancasters reporting attacks. All survived, however, and the gunners of Flying Officer M R Good's Lancaster shot down one of their attackers – a Messerschmitt Bf 109.

A most unusual incident occurred at this time when a Lancaster was attacked by a night fighter during one of 467 Squadron's Berlin attacks. Possibly after running out of ammunition, the German pilot ceased his attack and positioned his fighter near the Lancaster's tail fins where its mid upper gunner, Flying Officer H Bentley, could not fire on him. After maintaining formation for a short time, the German pilot waved a "cheerio" to the frustrated Australian gunner and disappeared into the night sky!

467 Squadron also participated in the raid on the German secret weapon establishment at Peenemunde on August 17/18 1943, where enemy opposition was especially severe. One Lancaster, piloted by Warrant Officer W L Wilson, was badly damaged by a night fighter over the target, however, the bomber's tail gunner succeeded in shooting the fighter down, despite severe damage to his turret. Operating at an intensive pace, 467 Squadron moved to Waddington – its best remembered base – in November 1943.

To cover the landing of Allied forces in France, 467 Squadron operated in strength on the night of June 5/6 1944, bombing coastal batteries on the invasion beaches. After the bridgehead had been established, operations continued, sometimes by day and, at times, in direct support of the advancing ground forces. On the night of March 23/24 1945, 467 Squadron assisted the advancing armies in an almost bloodless crossing of the Rhine River by attacking and destroying the enemy-occupied town of Wesel on the crossing point.

The Dortmund-Ems Canal, a vital German transport link, had come in for several attacks by 467 Squadron during 1944 and these continued into the new year. Always well defended, losses in these attacks were usually heavy and it was not a popular locality for the crews to "visit". On March 3/4 1945, for instance, no less than three of the 15 Lancasters despatched were lost to night fighters which had followed the bombers back to England. Some crews actually reported being attacked in the Waddington circuit!

In an earlier attack on the night of January 1, Flying Officer M G Bashe's Lancaster was hit by anti-aircraft fire three times on its run up to the target and another three times shortly after releasing its bomb load. With only two engines still functioning and other major damage, the pilot kept his crippled bomber in the air until it reached Dutch territory where the crew baled out. All seven airmen evaded capture and returned to England unharmed.

In a daylight attack against Hamburg on April 9, the bomber stream in which 467 Squadron was

(right) A famous Lancaster. 467 Squadron's "S for Sugar" celebrates 100 missions with its hard working ground staff. "S for Sugar" is preserved at the RAF Museum, Hendon.

flying, was attacked by German Messerschmitt Me 262 jet fighters. The Lancaster's piston engined fighter escorts could not cope with these high speed attacks and several bombers were lost, although none were from the Australian squadron.

With the European war almost concluded, some attacks were made against Czechoslovakian and Norwegian targets. Even before hostilities ceased in May, the Lancasters began the task of ferrying liberated POWs to England.

Selected as part of Tiger Force, the proposed RAF bomber group to operate against the Japanese in South East Asia, 467 Squadron moved to Metheringham in June to prepare for its new role, however, the collapse of Japan in August made these plans redundant and the unit disbanded on September 30 1945.

Lancaster R5868 'S for Sugar' survives today on display at the RAF Museum, Hendon, the only 467 Squadron Lancaster not scrapped after the war.

Formed: *November 7 1942*
Aircraft: *Lancaster*
Locations: *Scampton: November 1942*
Bottesford: November 1942 – November 1943
Waddington: November 1943 – June 1945
Metherington: June 1945 – September 1945
Disbanded: *September 30 1945*

SURVEY SQUADRON

Equipped with Australian built Mosquitos and commanded by Squadron Leader R M Green, Survey Flight was given squadron status on November 1 1946. Based at Fairbairn, the unit undertook photo survey tasks, both for the Defence Forces and the Australian Government prior to being redesignated 87 Squadron on March 8 1948.

Formed: *November 1 1946*
Squadron Code: *SU*
Aircraft: *Mosquito*
Location: *Fairbairn: Nov 1946 – March 1948*
Disbanded: *March 8 1948*

TARGET TOWING AND SPECIAL DUTIES SQUADRON

In September 1947 the Target Towing and Special Duties Flight at Richmond was given squadron status.

Commanded by Squadron Leader C A Greenwood, the unit was equipped with a variety of aircraft including Beaufighters, Mustangs, Ansons and single examples of the Wirraway, Beaufort and Dakota. It conducted a variety of tasks aside from its target towing role. These included army and naval co-operation duties, civil aid tasks and supporting CSIRO experiments. On March 8 1948 the unit was retitled 30 Squadron.

Formed: *September 1947*
Aircraft: *Beaufighter, Mustang, Anson, Dakota, Wirraway, Beaufort*
Location: *Richmond: Sept 1947 – March 1948*
Disbanded: *March 8 1948*

RAAF TRANSPORT FLIGHT VIETNAM

Hastily formed with three DHC-4A Caribou transports en route to Australia on their delivery flight, the RAAF Transport Flight Vietnam (RTFV) came into existence at Butterworth on July 20 1964.

Under the command of Squadron Leader C Sugden, the unit moved to Vung Tau, South Vietnam, on August 8 1964 and flew its first transport operations four days later. The RTFV's role was to provide transport support for South Vietnamese, American and "Free World" forces which were then facing the possibility of defeat against Communist forces.

Tasks included the transport of servicemen and civilians, freight and even livestock. South Vietnamese paratroops were regularly given jump training and some night time flare dropping operations in support of ground forces in contact with the enemy were conducted.

Typically, the RTFV operated throughout the length and breadth of South Vietnam. The Caribou were frequently subjected to ground fire but the precautionary tactic of tightly spiralling into and out of airfields significantly reduced this hazard. Nonetheless aircraft were hit by small arms fire and some crewmen and passengers were wounded.

Notwithstanding enemy action, which occasionally included mortar attacks against Vung Tau, and dangerous tropical flying conditions, the major hazard facing the RTFV was the condition of many of the landing strips the aircraft had to operate from. Frequently these were in very poor condition and a number of Caribou were damaged or destroyed while utilising these fields. Fortunately, the crews escaped injury in these incidents.

On August 29 1964 an additional three Caribou arrived at Vung Tau and a seventh and final aircraft was delivered in May 1965.

By this time the rugged Caribou and their skilful aircrews and dedicated ground staff quickly established themselves as the most efficient tactical transport unit to operate in Vietnam. So professional was the unit that US Army statisticians discovered that, while flying only 1% of the total transport missions in Vietnam, the small Australian unit was delivering no less that 7% of the total freight.

This phenomenal effort was a direct result of the aircrews doing everything possible to fly with a complete load. Additionally, the ground staff always ensured that the maximum number of Caribou were available for operations each day. Notwithstanding the arrival of several US Army efficiency experts at Vung Tau who observed RTFVs operations, the Americans and South Vietnamese were never able to match the Australian rate of effort on an aircraft to aircraft basis.

The RAAF Transport Flight Vietnam was redesignated 35 Squadron on June 1 1966.

Formed: *July 20 1964*
Squadron Code: *Nil*
Aircraft: *Caribou*
Locations: *Butterworth: July – August 1964*
Vung Tau: August 1964 – July 1966
Disbanded: *June 1 1966*

THE

ROYAL
AUSTRALIAN NAVY
FLEET AIR ARM

In 1911 the Australian Naval Forces were retitled the Royal Australian Navy (RAN). Just a few years later during World War I, RAN cruisers began experiments with shipborne aircraft.

With the formation of the RAAF in 1921, plans to form a Fleet Air Arm were shelved and, although the RAN did purchase a locally built seaplane carrier HMAS *Albatross* in 1928. However, the aircraft it operated were owned by the RAAF, flown by RAAF pilots and maintained by RAAF ground staff. *Albatross* was disposed of prior to World War II, and during that conflict amphibious aircraft were embarked in the RAN's cruisers to fulfil a variety of roles, the most important being reconnaissance and spotting for the ship's guns. Again, the aircraft were controlled by the ship's captain when embarked, but were RAAF property and were flown and maintained by air force personnel.

Events during the Pacific war forcibly displayed the value of naval air power and this lesson was not lost on the RAN which, almost as soon as hostilities had ended, laid plans to dramatically expand its naval aviation operations.

By 1948, these plans had become reality and on August 28 of that year, the Fleet Arm Arm's 20th Carrier Air Group commissioned at the Royal Navy's Eglinton Air Station in the UK. Equipped with British built Hawker Sea Furies and Fairey Fireflies, the 20th Carrier Air Group was joined in 1950 by the 21st Carrier Air Group, whose squadrons were also equipped with Sea Furies and Fireflies.

The young Fleet Air Arm "won its spurs" during the bitter Korean War when the light fleet carrier HMAS *Sydney* served in Korean waters from September 1951 to January 1952. During its participation in the conflict, *Sydney's* Sea Fury and Firefly squadrons flew 2366 sorties against enemy targets.

HMAS *Sydney* was paid off as an aircraft carrier in early 1955, while a loaned Royal Navy aircraft carrier, HMAS *Vengeance*, was operated between November 1952 and May 1955.

The RAN's third aircraft carrier, HMAS *Melbourne*, like *Sydney* and *Vengeance*, was a British designed and built light fleet carrier. *Melbourne* commissioned in the UK during October 1955 and arrived in Australia the following year carrying the RAN's new de Havilland Sea Venom jet fighters and Fairey Gannet anti-submarine warfare aircraft. The activities of the RAN's aircraft carriers were supported from the Fleet Air Arm's shore establishment at Nowra – HMAS *Albatross*. For a time, naval airfields at Schofields and Jervis Bay were also maintained.

In 1967, *Melbourne* bought the Fleet Air Arm's next generation of aircraft – Douglas A-4 Skyhawk fighter bombers and Grumman S-2 Tracker anti-submarine aircraft – home from the United States. Westland Wessex, and later Westland Sea King anti-submarine helicopters were also operated before *Melbourne* decommissioned in June 1982.

Plans to purchase a new carrier for the fleet lapsed in the early 1980s, forcing massive changes upon the Fleet Air Arm. All fixed wing squadrons were progressively disbanded and today the Fleet Air Arm exists solely as a rotary wing force (with the exception of two HS.748 electronic warfare aircraft). Anti-submarine and utility helicopters remain at sea on the RAN's FFG frigates, survey vessels and supply ships and will also be deployed on the new Anzac frigates as these locally built ships become operational.

The Fleet Air Arm continues to provide valuable support for the RAN's surface forces as evidenced by its effective operation of Sikorsky Seahawk and Aerospatiale Squirrel helicopters from Australian warships during the 1991 Gulf War. With its potent Seahawk force, the Fleet Air Arm greatly enhances the Royal Australian Navy's ability to detect and destroy hostile submarines and surface forces in Australia's region of direct military interest.

723 SQUADRON

Commanded by Lieutenant J A Gledhill and equipped with Sea Furies, one Dakota, one Sea Otter and a Wirraway, 723 Squadron commissioned at Nowra on April 7 1952.

The new unit's roles included communications flying, air sea rescue and co-operating with the Australian Joint Anti-Submarine warfare School. At times, 723 Squadron's aircraft were also used to provide refresher flying for naval aircrew not posted to flying squadrons.

Prior to decommissioning in October 1956, 723 Squadron had acquired Fireflies, Bristol Sycamore helicopters and its first and only jet aircraft – the de Havilland Vampire. The unit, and especially its helicopters, had also seen considerable service conducting civil aid tasks such as medical evacuations, rescues and flood relief work.

723 Squadron recommissioned at Nowra on February 18 1957 with Sycamores, Fireflies and Austers. The units main roles were helicopter pilot training and target towing. In November 1962 all fixed wing aircraft were transferred to

724 Squadron and in April 1963 the first of two Westland Scout helicopters for use on the survey vessel, HMAS *Moresby*, was delivered. While *Moresby* carried out its unglamorous but important survey duties in Australian and South West Pacific waters, one Scout along with its aircrew and a maintenance detachment were embarked as part of the ship's complement. In May 1964, Iroquois helicopters began to replace the Sycamores and were also used for helicopter pilot training.

From 1967, 723 Squadron provided operational training for the Royal Australian Navy Helicopter Flight Vietnam (RANHFV) which formed for active service with a United States Army Assault Helicopter Company. Although the RANHFV operated independently from 723 Squadron after arriving in Vietnam, it remained a sub unit of 723 Squadron. (The RANHFV's activities in Vietnam are dealt with separately.)

The Scout helicopters were replaced on board *Moresby* by Bell Kiowas in 1973 and by 1977 Wessexes had been added to the unit's fleet. During 1984, however, these aircraft were transferred to other units while French built Aerospatiale Squirrels were received. These new helicopters in turn replaced the ageing Iroquois. For a time Wessex operations recommenced when, in June 1987, the unit took over the helicopters of 816 Squadron.

During 1990, Squirrel detachments embarked in FFG class frigates were involved in a United Nations sponsored blockade of Iraq following that country's invasion of Kuwait. During this period the Squirrels conducted surveillance flights and delivered boarding parties onto merchant vessels. In 1991, United Nations forces commenced combat operations against Iraq and in addition to the activities already mentioned, the detachments conducted search and rescue tasks and surveillance patrols for mines set adrift by Iraqi forces. For a considerable period after the cessation of hostilities the FFGs and their

A Squirrel of 723 Squadron over the bow of an FFG.

Squirrel detachments remained overseas to support a continuing United Nations blockade.

Aside from the Kiowa and Squirrel helicopters, 723 Squadron also operates the Navy's two HS.748s. These unique aircraft were modified in the late 1970s to simulate an electronic warfare (EW) environment for training exercises.

Having gained valuable operational experience in the Persian Gulf, today 723 Squadron continues to operate in its fleet support role and, in addition, participates in numerous civil aid tasks in support of the Australian public.

***

Commissioned: April 7 1952
Aircraft: Auster, Dakota, Sea Otter, Wirraway, Sea Fury, Firefly, Sycamore, Vampire, Scout, Kiowa, Iroquois, Wessex, Squirrel, HS.748
Shore Station: Nowra: April 1952 – Oct 1956
Decommissioned:
Nowra: February 1957 – present
Embarked aboard: HMAS Moresby, HMAS Tobruk, HMAS Adelaide, HMAS Success, HMAS Sydney, HMAS Canberra, HMAS Darwin, HMAS Newcastle, HMAS Stalwart, HMAS Melbourne, HMAS Jervis Bay

***

724 SQUADRON

724 Squadron commissioned at Nowra on June 1 1955 under the command of Lieutenant Commander P R Dall. Equipped with Wirraway, Sea Fury, Firefly and Vampire aircraft, the unit's role was that of conversion training.

Sycamore helicopters were added to 724

Squadron's establishment in October 1956 while all the fixed wing aircraft except the Vampires were transferred to other units. Further changes later occurred when Sea Venoms and Gannets were bought on charge. These new aircraft bought with them a new role and 724 Squadron

724 Squadron operated Skyhawks in the training role. This aircraft is a TA-4 two seater.

found itself providing Sea Venom and Gannet operational training for front line squadrons embarked aboard HMAS *Melbourne*.

The disturbed nature of 724 Squadron's activities continued in 1957 when, for a period, it had no aircraft at all and very few aircrew prior to being redesignated a jet conversion and all weather fighter training squadron. With a small allocation of Sea Venoms, 724 Squadron formed its own aerobatic team, The Ramjets.

With the absorption of 725 Squadron in June 1961, 724 Squadron's collection of aircraft grew and encompassed Sea Venoms, Gannets, Vampires, Sea Furys, Dakotas and Austers. The unit's role was then altered to encompass all weather fighter, anti-submarine and operational flying training. 724 Squadron was also required to conduct general fleet support tasks, trials and communication work. In June 1963, 805 Squadron was absorbed while in November all the fixed wing aircraft of 723 Squadron were taken on charge.

By December 1968, 724 Squadron was fully equipped with jets operating a mix of A-4G Skyhawks, Vampires and Venoms which were used for conversion training. By now, in the twilight of their useful lives, the Venoms at this stage were also used as target tugs. The later delivery of Italian designed/Australian built MB-326H Macchi jet trainers resulted in the retirement of the Vampires and Venoms.

Despite 724 Squadron's rating as a second line squadron, the unit did participate in fleet exercises and on some occasions used its Skyhawks to provide close air support to ground troops during large scale ground manoeuvres. Perhaps remembering the days of the Ramjets, 724 Squadron also formed a Skyhawk aerobatic team – the Checkmates.

Until disbanding at Nowra on August 31 1984, 724 Squadron concentrated on both air-to-air and air-to-ground weapons training, photo reconnaissance tasks and target towing.

Commissioned: *June 1 1955*
Aircraft: *Sea Fury, Firefly, Wirraway, Vampire, Sea Venom, Gannet, Sycamore, Auster, Dakota, Skyhawk, Macchi*
Shore Station: *Nowra: June 1955 – Aug 1984*
Decommissioned: *August 31 1984*

---

725 SQUADRON

Under the command of Lieutenant Commander K C Barnett, 725 Squadron commissioned at Nowra on January 13 1958. Initially allocated a fleet requirements role, the unit was redesignated an anti-submarine training squadron in May 1959. 725 Squadron operated a variety of aircraft which included Gannet AS.1s, Fireflies, Sea Fury FB11s, a Dakota and an Auster. Prior to being decommissioned on May 31 1961, it was also allocated a number of Venom FAW.53s.

Recommissioning as an operational anti-submarine helicopter squadron on November 1 1962, 725 Squadron equipped with Westland Wessex anti-submarine helicopters. Four of these were attached to the fast troop transport and former aircraft carrier, HMAS *Sydney*, to provide anti-submarine cover to that ship during its many voyages to Vietnam. In the anti-submarine role, the Wessexes utilised their dipping sonar while an offensive armament comprised torpedoes and depth charges. 725 Squadron decommissioned at Nowra on December 27 1975.

Commissioned: *January 13 1958*
Aircraft: *Dakota, Auster, Gannet, Sea Fury, Firefly, Wessex*
Shore Stations: *Nowra: Jan 1958 – May 1961*
Decommissioned:
Nowra: November 1962 – December 1975
Embarked aboard: *HMAS Sydney*
Decommissioned: *December 27 1975*

---

805 SQUADRON

Initially, equipped with Sea Furies, 805 Squadron was the Fleet Air Arm's first fighter squadron and commissioned at the Royal Naval Air Station Eglinton in Northern Ireland on August 28 1948. Commanded by Lieutenant Commander P E Bailey, the unit embarked in *Sydney* and arrived in Australia during February 1949.

After three years of peacetime flying, 805 Squadron sailed for Korea in late 1951 and commenced combat operations against North Korean and Chinese forces on October 5. The Sea Fury squadron flew at an intensive rate and participated in armed reconnaissance and ground attack missions – including close-in attacks in support of friendly ground forces – and attacks on North Korean watercraft. 805 Squadron pilots also at times directed naval gunfire and flew combat air patrols over *Sydney's* Task Force.

Highlights of 805 Squadron's tour included rocket and gunnery attacks on 1000 communist troops on October 11. The Sea Furies of 805 and 808 Squadrons inflicted an estimated 200 casualties on the enemy before the attack was broken off. It was on this day that *Sydney's* Carrier Air Group flew a record 89 sorties.

A few days later, 805 Squadron personnel endured Cyclone Ruth. During a night of sheer terror, *Sydney* pitched violently from side to side at extreme angles and a number of electrical fires had to be extinguished. Considerable damage was done to the ship, several aircraft were damaged and one Firefly was washed overboard.

805 Squadrons operations, invariably conducted at low level, drew heavy and accurate anti-aircraft fire and many Sea Furies were damaged. Ironically, of *Sydney's* three squadrons, 805 Squadron suffered all the fatalities. On November 5 1951, Lieutenant K E Clarkson's Sea Fury disintegrated after being hit by anti-aircraft fire. This officer was a highly experienced former RAAF pilot who had been decorated with the Distinguished Flying Medal in World War II.

On December 7 Sub Lieutenant R Sinclair was killed after striking the tailplane of his stricken fighter while attempting to bale out near Chinnampo. 805 Squadron's last loss occurred on January 2 1952 when Sub Lieutenant R J Coleman's Sea Fury disappeared over the Yellow Sea, no trace of the pilot ever being found.

Several other pilots had lucky escapes after making forced landings – perhaps one of the most unusual occurring on October 26 1951 when, after crash landing on a mud flat, Sub Lieutenant N W Knapstein salvaged what he could from his wrecked Sea Fury and then sold the rest of the aircraft to Korean villagers before being rescued by a Royal Navy vessel. In another incident on December 13, Lieutenant Commander W C Bowles was shot down and baled out over the sea from where he was rescued by a passing junk and returned to *Sydney* three days later.

Sydney sailed for Australia on January 27 1952, allowing the ship's complement to enjoy some well deserved shore leave. 805 Squadron had not seen the last of Korea however, and the unit, again embarked in *Sydney*, served in Korean waters for a period after the ceasefire on garrison duty.

Returning to Australia again, 805 Squadron undertook routine duties for some time, eventually becoming the last piston engined fighter squadron in the Fleet Air Arm. Prior to this, on August 30 1955, the unit had received considerable media attention when two of its pilots shot down an empty civilian Auster light aircraft (following a failed attempt by the RAAF) which had managed to takeoff from Bankstown without the assistance of its pilot.

805 Squadron decommissioned at Nowra on March 26 1958, only to recommission with de Havilland Sea Venom FAW.53 all weather jet fighters one week later, on March 31. With these aircraft, the unit embarked in *Melbourne* and participated in numerous exercises both in Australian waters and in the Pacific region until again decommissioning on June 30 1963.

On January 10 1968, 805 Squadron recommissioned with McDonnell Douglas A-4G Skyhawk

Sea Furies of 805 Squadron served in Korea.

fighter bombers and again embarked in *Melbourne*, operating in the air defence, anti-shipping and army support roles. The rugged and easy to maintain Skyhawks carried a variety of weapons to fulfil their tasks including AIM-9B Sidewinder infrared air-to-air missiles, high explosive bombs and air-to-ground rockets.

805 Squadron continued to operate from *Melbourne* during the 1970s. *Melbourne* operated mainly in the Pacific region and travelled as far afield as Hawaii and Japan where 805 Squadron participated in a variety of exercises which allowed it to hone its combat skills. These exercises involved the armed forces of the USA and many Asian countries.

Unfortunately, the Australian Government's decision to end fixed wing Fleet Air Arm operations forced the decommissioning of one of the Fleet Air Arm's finest squadrons at Nowra on July 2 1982.

Commissioned: August 28 1948
Aircraft: Sea Fury, Sea Venom, Skyhawk
Shore Stations: Eglinton: Aug 1948 – Feb 1949
Nowra: February 1949 – March 1958
Decommissioned:
Nowra: March 1958 – June 1963
Decommissioned:
Nowra: January 1968 – July 1982
Embarked aboard: HMAS Sydney, HMAS Vengeance, HMAS Melbourne
Decommissioned: July 2 1982

808 SQUADRON

808 Squadron commissioned at St Merryn in the UK on April 25 1950. Equipped with Hawker Sea Fury FB.11s, the new fighter squadron, commanded by Lieutenant Commander J L Appleby, worked up to operational status prior to embarking in HMAS *Sydney* and sailing for Australia in December.

By mid 1951, the personnel of 808 Squadron knew that they would be committed to support United Nations forces in Korea later in the year. Operational training was stepped up and by the first week of October *Sydney* was in Korean waters.

Combat flying commenced on October 5 when ground attack missions were flown despite the

Sea Venom landing on *Melbourne*.

prevalent poor weather conditions which limited their effectiveness. Utilising an armament of air-to-ground high explosive rockets and 20mm cannons, 808 Squadron's tasks included ground attack operations, including the provision of close air support to friendly ground forces, armed reconnaissance, spotting for ships guns and anti-shipping strikes – principally against North Korean small craft. Additionally two fighters had to remain over *Sydney's* Task Force throughout the hours of daylight to protect the ships in the event of a surprise air attack.

Just days after commencing operations in Korea, *Sydney's* aircraft created a record by flying no less than 89 sorties. In the course of this day's flying (October 11), both Sea Fury squadrons flew intensively and on one mission located some 1000 North Korean troops who were strafed and rocketed. A conservative estimate later put the communist losses at two hundred killed.

Operations continued at an intensive rate throughout *Sydney's* tour in Korean waters and enemy ground fire repeatedly damaged the Australian Sea Furies, sometimes seriously. In one incident on October 25, Lieutenant C M Wheatly was hit by ground fire and forced to bale out over the sea. Fortunately, an American air-sea rescue amphibian was able to land and rescue the Australian, who was later returned to his ship.

In another incident on January 5 1952, Lieutenant P Goldrick was wounded in the arm by anti-aircraft fire but managed to return to *Sydney* where he made a safe landing. This pilot was destined to be both 808 Squadron's only casualty of the war.

The Sea Furies continued to achieve valuable results and it became noticeable that the enemy showed a marked reluctance to move trains or road traffic in *Sydney's* area of operations by day. Finally, in the depths of a Korean winter, 808 Squadron flew its final missions of the Korean War on January 25 1952 and sailed for Australia a few days later.

In October 1953, the unit embarked in HMAS *Vengeance* and continued carrier borne operations until decommissioning at Nowra on October 5 1954. 808 Squadron recommissioned as an all weather fighter squadron at the Royal Naval Air Station Yeovilton on August 23 1955. Equipped with Sea Venom FAW.53s, the squadron participated in the new carrier *Melbourne's* flying trials before proceeding to Australia in March 1956.

Over the next few years the radar equipped Venoms participated in numerous exercises – particularly in South East Asia – and the unit attained a reputation as being a particularly effective fighter squadron giving a good account of itself even when flying against aircraft possessing superior performance. 808 Squadron remained operational until December 1 1958 when it decommissioned at Nowra.

Commissioned: *April 25 1950*
Aircraft: *Sea Fury, Sea Venom*
Shore Stations: *St Merryn: April – Aug 1950*
Nowra: August 1950 – October 1954
Decommissioned:
Yeovilton: August 1955 – February 1956
Nowra: February 1956 – December 1958
Embarked aboard: *HMAS Sydney, HMAS Vengeance, HMAS Melbourne*
Decommissioned: *December 1 1958*

---

816 SQUADRON

816 Squadron commissioned with Fireflies at the Royal Naval Air Station at Eglinton, Northern Ireland, on August 28 1948.

As with the other Australian Fleet Air Arm squadrons formed in the UK at this time, a number of Royal Navy pilots were attached to the unit until the Australians obtained a suitable level of carrier operating experience. This practice was prevalent throughout the Fleet Air Arm and Royal Navy officers and ratings would be a common sight on Australian carriers and shore stations for several years. The new Firefly AS.5 squadron, commanded by Lieutenant Commander C R J Coxon, along with 805 Squadron, formed the Fleet Air Arm's 20th Carrier Air Group and, embarked aboard HMAS *Sydney*, arrived in Australian waters during May 1949.

Have attained a suitable level of operational proficiency in the United Kingdom, and after conducting further training at Nowra, 816 Squadron embarked in *Sydney* and served in Korean waters for several months after the ceasefire.

After a further period, 816

Squadron embarked in *Vengeance*, and decommissioned at Nowra on April 27 1955. This was a temporary measure, however, as four months later on August 15, the unit recommissioned in the United Kingdom.

With its new Fairey Gannet AS.1 anti-submarine aircraft, 816 Squadron commenced its own training programme from the Royal Naval Air Station at Culdrose, Cornwall, and as this progressed, participated in the flying trials for the Fleet's new carrier, HMAS *Melbourne*. Embarked in this ship, 816 Squadron arrived in Australia during May 1956.

Gannets of 816 Squadron (pictured) operated off HMAS *Melbourne*.

Before operating Seahawks, 816 Squadron last operated the Wessex (foreground). Both are pictured at Nowra. (Jim Thorn)

The Gannets remained in service for several years and proved a most useful aircraft. In the anti-submarine role, the aircraft's search radar was backed up by an impressive armament of homing torpedoes, depth charges and rockets. As a secondary role, the Gannets were also used to search and/or shadow shipping. In July 1964 a flight of Sea Venom all weather fighters were added to the unit's inventory. Plans were in hand to obtain a new generation of aircraft however, and preparatory to re-equipping with Grumman S-2E Trackers, 816 Squadron decommissioned at Nowra on August 25 1967.

Recommissioning on January 10 1968, the unit's Trackers operated from *Melbourne* and its shore station at Nowra. The Trackers proved to be excellent anti-submarine aircraft and utilised active and passive sonobuoys to locate submarines. They carried a range of weapons including anti-submarine torpedoes, depth charges, bombs and rockets. With its two engines, the aircraft gave the crews confidence when flying on long range over water flights.

816 Squadron suffered a devastating setback on December 4 1976 when most of its aircraft were burnt in a major hangar fire at Nowra. Flying was severely curtailed for some months until replacement S-2G Trackers could be delivered from the United States. With these former US Navy Trackers, 816 Squadron re-embarked in

Melbourne and participated in numerous exercises throughout the Pacific region and in Australian waters. Other activities undertaken included searches for missing vessels and occasional surveillance flights searching for refugee boats and illegal fishing in Australian waters. 816 Squadron decommissioned at Nowra on July 2 1982.

Equipped with Westland Wessex Mk.31B helicopters, 816 Squadron recommissioned on February 7 1984 and operated in the specialist army support role, often maintaining detachments on the landing ship HMAS *Tobruk*, for the three and a half years until decommissioning in June 1987.

816 Squadron recommissioned yet again on July 23 1992. Equipped with the advanced Seahawk S-70B Seahawk anti-submarine helicopters, the unit now operates detachments from FFG class frigates. 816 Squadron today is the Fleet Air Arm's most combat capable unit and represents a vital element in the Royal Australian Navy's anti-submarine defence.

Commissioned: *August 28 1948*
Aircraft: *Firefly, Gannet, Sea Venom, Tracker, Wessex, Seahawk*
Shore Stations: *Eglinton: Aug 1948 – Feb 1949*
Nowra: February 1949 – April 1955
Decommissioned:
Culdrose: August 1955 – February 1956
Nowra: February 1956 – August 1967
Decommissioned:
Nowra: January 1968 – July 1982
Decommissioned:
Nowra: February 1984 – June 1987
Decommissioned:
Nowra: July 1992 – present
Embarked aboard: *HMAS Sydney, HMAS Vengeance, HMAS Melbourne, HMAS Tobruk, HMAS Adelaide, HMAS Newcastle, HMAS Canberra, HMAS Darwin, HMAS Sydney, HMAS Melbourne, HMAS Stalwart*

---

817 SQUADRON

817 Squadron commissioned on April 25 1950 at St Merryn in the United Kingdom.

With 808 Squadron, the new Firefly equipped anti-submarine unit, which was under the command of Lieutenant Commander R B Lunberg, formed the 21st Carrier Air Group.

After arriving in Australia during December 1950 on board HMAS *Sydney*, the Fireflies par-

ticipated in numerous exercises before *Sydney* deployed to Korea to support United Nations forces. Prior to the deployment, 817 Squadron swapped its AS.6 Fireflies, for the earlier Mk.5 which possessed cannon armament which the AS.6 did not require in its highly specialised anti-submarine role.

Arriving in Korean waters during the first week

of October 1951, *Sydney* began launching air strikes on October 5. 817 Squadron's share in this first days flying, which was limited due to poor weather, was eight sorties – four Fireflies attacked a bridge near Songwha while another four conducted anti-submarine patrols. Next day, the unit had its first major success when its crews damaged railway lines and destroyed spans of a road bridge at Chinnampo. In a short space of time, 817 Squadron became remarkably proficient in the bridge busting role and on several occasions during *Sydney's* Korean patrols, the Fireflies destroyed every bridge in the carrier's area of operations.

817 Squadron's main role for the duration of the conflict was that of rail/road interdiction. For this task high explosive bombs and the Fireflies 20mm cannons were utilised in low level attacks against heavily defended targets behind enemy lines. Anti-aircraft fire was particularly effective under these circumstances and, although never turned from their targets, some Fireflies were lost and others damaged. Other roles undertaken included flying anti-submarine patrols around *Sydney's* task force, and spotting for ships main armaments during shore bombardments.

On October 11, 817 Squadron flew intensively and played its part in *Sydney's* Carrier Air Group flying a record 89 sorties. To achieve this feat, the carrier's deck teams worked constantly, re-fuelling and rearming aircraft for additional strikes as they landed and removing damaged or unserviceable aircraft to the hangars.

817 Squadron's personnel, along with the rest of *Sydney's* company, had the misfortune of enduring Cyclone Ruth near the Japanese coast on October 14. The carrier suffered considerable damage during a nightmare night. Some personnel were injured, although not seriously, however, one Firefly was washed overboard in the mountainous seas and several others were damaged.

On October 26, a Firefly crewed by Sub Lieutenant N MacMillan and Chief Petty Officer P Hancox was badly damaged by ground fire and forced down behind enemy lines near Sariwon. Sea Furys from *Sydney* and Meteor jet fighters from the RAAF's 77 Squadron hastened to the scene and provided covering fire against nearby enemy troops. Despite the lateness of the day *Sydney's* American manned Sikorsky rescue helicopter was despatched to the scene and plucked the two Australians from enemy territory. As it was almost dark and the small helicopter did not have night flying instrumentation, it diverted to Kimpo and returned to *Sydney* with its passengers the following day.

Over the succeeding weeks, 817 Squadron continued its difficult and dangerous operations. The onset of winter with the resulting icy conditions and subzero temperatures made aircraft handling on *Sydney's* windswept deck a nightmare. Frostbite became an occupational hazard for the sailors and the aircrews ran the additional risk of freezing to death if forced down in the sea.

817 Squadron had a remarkably successful day of operations on January 20 1952 destroying four railway bridges and damaging a fifth. The unit flew its last missions of the Korean War four days later, after which *Sydney* sailed for Australia. During its participation in the Korean War, 817 Squadron flew 743 operational sorties for the loss of three aircraft as well as damage to many others. Fortunately no personnel were killed during the tour.

After returning to Australia, 817 Squadron re-equipped with its AS.6 Fireflies and participated in various exercises and later, the British atom bomb tests at Monte Bello. In March 1953, the unit sailed for the United Kingdom for the Coronation of Queen Elizabeth. Perhaps the highlight of this visit was the participation of eight Australian Fireflies in a 17 squadron flypast for the new Queen at Spithead in April.

817 Squadron embarked in HMAS *Vengeance* during October 1953 and remained operational in that ship for 11 months. On April 27 1955, the unit decommissioned at Nowra only to recommission at Culdrose in Cornwall during August with Fairey Gannet anti-submarine aircraft. 817 Squadron remained in the United Kingdom working up to operational proficiency and then participating in the flying trials of the newly commissioned carrier *Melbourne*, prior to arriving in Australia during February 1956. Embarked in *Melbourne*, the Gannets participated in numerous exercises prior to decommissioning yet again on August 18 1958.

A Sea King anti-submarine helicopter of 817 Squadron.

On July 18 1963, 817 Squadron again came into existence when it recommissioned at Nowra with British built Westland Wessex 31A anti-submarine helicopters. The helicopters operated from *Melbourne* and participated in numerous deployments during the 1960s. The importance of these exercises was underscored during Indonesia's Confrontation with Malaysia – a period when Australian troops and naval forces were, at times, at combat status against Indonesian forces. A further commitment commenced later when four Wessex helicopters operated from the fast transport *Sydney*, providing that ship with anti-submarine protection during its journeys to and from South Vietnam.

As a secondary role, 817 Squadron provided guard helicopters during the launching and recovery of *Melbourne's* fixed wing aircraft. Carrying a diver, the helicopter would remain airborne to one side of the carrier and, in the event of a ditching during takeoff, quickly drop the diver at the scene to help free any trapped aircrew. The Wessex helos, when flying from Nowra, were also used for civil aid tasks such as search and rescue and medevac missions.

In early 1969, 817 Squadrons re-equipped with the more capable Wessex 31B and continued to provide anti-submarine protection for the Australian Fleet. Between September and December, Wessex were embarked aboard the British Royal Fleet Auxiliary RFA *Tidespring* – a unique experience in the history of the Australian Fleet Air Arm.

During 1975, the Wessex took on a secondary role when the unit received its first HAS.50 Sea King helicopters. The Westland built Sea Kings bought with them a quantum leap in technology and offered considerable advancements in the detection of submarines over the Wessex.

Since the disposal of the RAN's last aircraft carrier in 1982, 817 Squadron has operated permanently from Nowra as the Sea Kings are too large to be embarked in the RAN's FFG frigates which instead utilise Seahawks to provide their anti-submarine capability. Nonetheless, the Sea Kings have at times flown from some of the Fleet's support vessels specifically HMA ships *Success*, *Jervis Bay* (now retired) and *Tobruk*. Additionally, the unit conducts ships transfers, support tasks and civil aid activities. The Sea Kings will also operate off the two new Training and Helicopter Support Ships HMA ships *Kanimbla* and *Manoora* once they are refitted.

Commissioned: *April 25 1950*
Aircraft: *Firefly, Gannet, Wessex, Sea King*
Shore Stations: *St Merryn: April – June 1950*
Eglinton: June – July 1950
St Merryn: July – August 1950
Nowra: August 1950 – April 1955
Decommissioned:
Culdrose: August 1955 – February 1956
Nowra: February 1956 – August 1958
Decommissioned:
Nowra: July 1963 – present
Embarked aboard: *HMAS Sydney, HMAS Vengeance, HMAS Melbourne, RFA Tidespring, HMAS Success, HMAS Tobruk, HMAS Jervis Bay*

---

850 SQUADRON

Under the command of Lieutenant Commander R A Wilde, 850 Squadron commissioned at Nowra on January 12 1953.

Equipped with Hawker Sea Fury FB.11s, the new fighter squadron had only a brief existence and after achieving operational status embarked in *Sydney* for a post ceasefire patrol in Korean waters.

Between December 1953 and January 1954, 850 Squadron logged a month long shore detachment at Kai Tak, Hong Kong and after a short period embarked in *Vengeance* decommissioned on August 3 1954.

Commissioned: *January 12 1953*
Aircraft: *Sea Fury*
Shore Station: *Nowra: Jan 1953 – Aug 1954*
Embarked aboard: *HMAS Sydney, HMAS Vengeance*
Decommissioned: *August 3 1954.*

---

851 SQUADRON

Commanded by Lieutenant D Johns, 851 Squadron commissioned at Nowra on August 3 1954.

Equipped with Fireflies and Dakotas the new unit operated in a training role providing pilots and observers to the Fleet Air Arm's frontline Firefly squadrons. The Dakotas, aside from their observer training tasks, were also utilised for transport and communications duties and in this role sometimes carried VIPs and dignitaries which, on one occasion, included the Duke of Edinburgh and Lord Mountbatten.

With the planned reduction of aircraft carriers in the RAN from two to one in the mid 1950s, the need for 851 Squadron diminished and the unit decommissioned at Nowra on January 13 1958. Despite its allocated training role, it did get to sea briefly during March 1956 when it had embarked for two weeks in *Sydney*.

Equipped with S-2E Trackers and Dakotas, 851 Squadron recommissioned as a training and transport unit at Nowra on September 2 1968. With its mix of long range aircraft, 851 Squadron often travelled throughout Australia. In 1973, two HS.748s replaced the Dakotas.

Aside from its training and transport roles, 851 Squadron's Trackers were called upon to search for missing small craft and also, at times, to conduct

security patrols of Bass Strait oil facilities. Additionally, coastal and fishery patrols in northern Australia were carried out – often by detachments operating from Darwin and Broome. After absorbing 816 Squadron in July 1982, 851 Squadron itself decommissioned on August 31 1984.

Commissioned: *August 3 1954*
Aircraft: *Firefly, Dakota, Tracker, HS.748*
Shore Stations: *Nowra: Aug 1954 – Jan 1958*
Decommissioned:
Nowra: September 1968 – August 1984
Embarked aboard: *HMAS Sydney*
Decommissioned: *August 31 1984*

An 851 Squadron Tracker.

ROYAL AUSTRALIAN NAVY HELICOPTER FLIGHT VIETNAM

In July 1967, the Minister for the Navy announced that a detachment of naval aviators and aircraft maintainers would be sent to South Vietnam.

As the sailors would be flying and maintaining United States Army helicopters, a new unit, the Royal Australian Navy Helicopter Flight Vietnam (RANHFV) formed as a sub unit of 723 Squadron at Nowra.

The first contingent of the RANHFV comprising eight pilots, four observers, four aircrewmen and 24 maintainers and six non technical support staff commenced training in their new army support role before proceeding to Saigon via Qantas aircraft in two groups between October 16-18 1967.

From Saigon, the Australians, under the command of Lieutenant Commander N Ralph, travelled to Vung Tau where they were integrated in

their new unit, the 135th Assault Helicopter Company (135th AHC). Operating around 30 Bell UH-1H and UH-1C Iroquois helicopters the 135th AHC's role was to provide air mobility to American, Vietnamese and other "Free World" troops.

Within the unit, the Australians were placed in positions commensurate with their rank and experience. Lieutenant Commander Ralph, while retaining administrative command of the Australians, became second in command of the 135th AHC which was led by Lieutenant Colonel R Cory.

Two platoons of the 135th AHC flew lightly armed troop lift helicopters (slicks) while a third operated the unit's UH-1C Gunships. A maintenance platoon undertook the servicing and repair of the three flying platoons' helicopters.

Before commencing flying with the 135th AHC, the naval aviators were split up and attached to various helicopter companies around Vietnam for two weeks to gain experience and in the process many came under fire for the first time.

On reassembling at Vung Tau, the 135th AHC was declared fully operational and began combat flying with its RAN component on November 2. The unit operated throughout much of Vietnam and its large scale operations frequently met with heavy opposition from the Viet Cong. This was especially so during assault landings when the enemy was either in or near landing zones. Consequently, losses at times, were heavy.

The first Australian to be downed by enemy fire was Sub Lieutenant A A Casadio, whose gunship crash landed after being damaged by ground fire. On the ground the Australian pilot and his American crew were attacked by the Viet Cong. The helicopter's machine guns were detached and used to excellent effect on the advancing enemy, at least two of whom were killed before the crew were rescued by another helicopter.

In December the 135th AHC moved from its well established base at Vung Tau to Camp Blackhorse in Long Kanh Province. Flying commitments were not reduced during the move and, as the Company's aircrew were logging 140 flying hours per month – about a years' peacetime flying in Australia – the maintenance crews worked long hours to keep the Hueys airworthy while at the same time setting up new maintenance facilities at Blackhorse. It is significant to note that the maintenance personnel also regularly flew on operations to enable crewmen to have a break from their flying duties.

The naval component of the 135th AHC suffered its first casualties on January 12 when a Huey crashed near Baria. Leading Mechanic K J French and Naval Airman K R Wardle were both injured in the incident. On February 22, while leading a platoon which was tasked to extract South Vietnamese troops near Xuan Loc, the lead helicopter, flown by Lieutenant Commander P J Vickers, was hit by ground fire. One round penetrated the cabin, fatally wounding the Australian pilot.

On August 21, as the naval aviators' tour of duty was drawing to a close, a gunship captained by Sub Lieutenant Casadio was hit by a rocket propelled grenade near Blackhorse and crashed in flames. The entire crew, including Petty Officer O C Phillips, were killed.

135th AHC Hueys operating in the "Delta" area.

In early September, the First Contingent of the RANHFV was replaced by the Second Contingent. Commanded by Lieutenant Commander G R Rohrsheim, the newly arrived sailors quickly settled into their unfamiliar environment and were soon in action. Operations continued at an intensive pace and on October 23 no fewer than 11 Hueys were hit by ground fire in the Ben Tre area. Two were so badly damaged that they were later written off.

The 135th AHC again changed bases in November when it moved to Camp Bearcat, a large Thai Army base 25km south east of Saigon. The move to Bearcat meant that the unit would be operating almost solely in the Mekong Delta – an area of intense enemy activity.

On January 3 1969, Sub Lieutenant A J Huelin and his American crew were killed when their helicopter struck high voltage wires near Saigon. The Australians had many close shaves over the succeeding months but it was not until May 31 that another fatality was sustained. On this day, a gunship in which Leading Aircrewman N E Shipp was flying was hit by ground fire and crashed with the loss of all crew. Despite these casualties, the 135th AHC continued its effective operations and in an action soon afterwards a gunship captained by a navy pilot killed at least 30 enemy troops.

In September 1969, the Third Contingent of the RANHFV replaced the Second and, as intensive army support operations continued, the newly arrived Australians suffered their share of lucky escapes and misfortunes. Several Navy personnel were wounded and it was indeed fortunate that none were killed. One of the luckiest pilots perhaps was Lieutenant R Marum who, on January 22, was wounded in the leg by a bullet which, as it continued its upwards flight, tore through his hand before hitting him in the chest. No further injury was caused to the Australian, however, as he had wisely taken the precaution of wearing his "chicken plate" body armour.

The 135th AHC again relocated in September 1970, this time to Dong Tam in Dinh Tuong Province. On the heels of this move, the fourth, and final contingent of the RANHFV, arrived in Vietnam. Like their predecessors, the newly arrived Australians continued the dangerous pattern of missions typical of any US Army helicopter company in Vietnam.

On December 4 Lieutenant J C Buchannan effected a remarkable rescue of a disabled South Vietnamese patrol boat which was under mortar fire and drifting towards an enemy occupied shore line. Hooking his Huey's skids within the boats superstructure he towed the vessel away from the enemy, enabling the craft's crew to be rescued.

As part of Australia's disengagement from the conflict in Vietnam, the Fourth Contingent was withdrawn from the 135th AHC earlier than expected. Flying ceased on June 8 1971 and the RANHFV left Vietnam four days later.

During their time in South Vietnam, the four contingents of the RANHFV earned themselves a formidable reputation for bravery and professionalism second to none in the United States Army helicopter forces. Losses had been proportionally high but the naval personnel had demonstrated their adaptability and, in the process, earned their parent unit – 723 Squadron – the battle honour "Vietnam 1967 – 1971".

Formed: *October 1967*
Aircraft: *Iroquois*
Locations: *Vung Tau: October – December 1967*
Camp Blackhorse: Dec 1967 – November 1968
Camp Bearcat: November 1968 – Sept 1970
Dong Tam: September 1970 – June 1971
Disbanded: *June 1971*

THE

AUSTRALIAN ARMY AVIATION CORPS

The Australian Army Aviation Corps, the Army's "second air corps", began its evolution immediately after World War II when, in 1947, a Royal Australian Artillery Corps officer commenced pilot training in the United Kingdom.

World War II had repeatedly shown the necessity for artillery fire to be directed by pilots with a thorough understanding of the uses and limitations of artillery. Needless to say, none was more experienced in this specialised branch than the artilleryman himself.

Despite the lessons of the war, the Army was not immediately able to establish its own air corps due to RAAF and political opposition to the plan. Notwithstanding these difficulties, Army Aviation did slowly evolve. Several more Artillery Corps officers underwent pilot training in the United Kingdom and the RAAF itself belatedly commenced Army pilot training at Fairbairn from mid 1952 using Auster aircraft. Also at this time, Army officers commenced training at Point Cook as flying instructors.

In June 1953, 16 (AOP) Flight formed taking over the Austers of 3 Squadron. The new unit, while still part of the RAAF, had a large Army component. In the interim, Army pilots had been attached to an RAF AOP Flight in Korea to obtain combat experience. 1 Army Aviation Company formed at Bankstown during 1957 to supplement 16 (AOP) Flight, however this unit, comprising pilots only, who flew chartered civilian light aircraft, was of limited utility.

With an agreement finally reached between the RAAF and Army over light aircraft operations, the way was at last clear for the serious development of Army Aviation and, on December 1 1960, 16 Army Light Aircraft Squadron formed at Amberley. Equipped with Sioux helicopters and fixed wing Cessnas, the new unit was initially commanded by an RAAF officer and, despite the absorption of personnel from 16 (AOP) Flight and 1 Aviation Company, initially no Army technical ground staff were available and RAAF personnel carried out all maintenance tasks until soldiers could be trained to undertake these duties.

Since the formation of 16 Army Light Aircraft Squadron, the Army Aviation Corps has continued to expand. In the mid 1960s, small detachments were deployed to New Guinea and Malaysia (the later commitment being on active service against Indonesian infiltrators). In 1965, 161 Reconnaissance Flight deployed to South Vietnam and provided an exceptional degree of support, initially to the 173rd Airborne Brigade and later to the 1st Australian Task Force.

In the post-Vietnam era, the Australian Army Aviation Corps – which formally came into existence on July 1 1968 – has continued to increase in size and in the scope of its operations. In the late 1980s, the Army Aviation Corps accepted responsibility for battlefield helicopter operations from the RAAF. Today's Black Hawks, Iroquois, Kiowas, Squirrels and Nomads were joined by Chinooks in mid 1995 and possibly new armed attack/reconnaissance helicopters.

With this combination of aircraft and experienced personnel, the Australian Army Aviation Corps continues to provide effective battlefield support to the Australian Defence Force.

16 SQUADRON

16 Army Light Aircraft Squadron formed as a joint Army/RAAF unit at Amberley on December 1 1960.

Commanded by Wing Commander K V Robertson, its role was to support Army activities and at the same time to train Army pilots – and later technical ground staff – for a planned increase in Army Aviation strength over the next decade.

Equipped with Cessna 180 fixed wing aircraft and Bell Sioux helicopters, the unit made an early beginning to pilot training while simultaneously participating in numerous exercises. The latter took the helicopters and Cessnas away from Amberley regularly and they operated over much of Australia. Operationally, the unit's main role was to provide aerial reconnaissance, however, casualty evacuations and communication tasks were also undertaken.

From the beginning, 16 Squadron conducted numerous civil aid tasks such as undertaking search and rescue missions and medical evacuations. One unsuccessful search task took two Sioux as far afield as Merauke in Dutch New Guinea. Perhaps one of the most unusual incidents in this early period occurred on July 20 1961 when a Cessna piloted by Lieutenant D S Bell observed a tribal battle developing between Aboriginals at Warburton River Mission. Swooping down over the confrontation, the startled warriors bolted and, at least for the moment, forgot their differences in their haste to escape the fearsome Cessna!

In November 1962, 16 Squadron was directed to support United Nations activities in New Guinea. Again deploying to Merauke via RAAF Hercules, the Sioux ferried anti cholera vaccine and medical personnel to various villages. Despite the loss of one helicopter in a crash, this task was successfully completed and the detachment returned to Amberley late in December.

Over the next few years, pilot training and exercise commitments continued with few highlights. In February 1964, a Cessna deployed to various locations in New Guinea – the only major difference to earlier operations there being that the aircraft flew all the way there and back. In April, another overseas exercise occurred when a Cessna and Sioux were deployed to Ubon, Thailand.

On December 18 1964, 16 Squadron received its first Army Commanding Officer when Lieutenant Colonel W J Slocombe took command of the unit. This was more than a symbolic gesture as, from this date, control of the unit passed from the RAAF to the Army. December also saw the first permanent detachment to New Guinea commence.

On April 26 1966, 16 Army Light Aircraft Squadron was renamed 1st Divisional Aviation Regiment, a title amended to 1st Aviation Regiment early in 1967. Despite the name changes, unit activities continued with little change.

After relocating to Oakey, the 1st Aviation Regiment underwent a major organisational change on August 1 1972 when Headquarters

Cessna 180s, along with Sioux helicopters, comprised 16 Army Light Aircraft Squadron's initial equipment.

Army Aviation Centre, the School of Army Aviation and Army Aviation Centre Base Squadron, were all formed from unit resources. Later, with the formation of subordinate flying squadrons, the 1st Aviation Regiment shed its aircraft and became the controlling headquarters for reconnaissance and support operations for the Australian Army.

Formed: December 1 1960
Aircraft: Cessna 180, Sioux
Locations: Amberley: Dec 1960 – April 1966
Renamed 1st Divisional Aviation Regiment 26 April 1966

161 SQUADRON

Under the command of Major P D Lipscombe, 161 Reconnaissance Flight was raised at Amberley in June 1965 and, after a brief but intensive work up period, arrived at Bien Hoa, South Vietnam, between September 28 and 29.

The small unit, equipped with two Bell Sioux light observation helicopters and two Cessna 180 fixed wing aircraft, was to provide tactical reconnaissance for the Australian 1st Battalion Group attached to the United States Army's 173rd Airborne Brigade. Due to a shortage of Army trained maintenance personnel, initially many of the maintenance support personnel were RAAF members.

161 Flight soon proved more versatile than first envisaged and began to undertake additional tasks, both for Australian and American troops. These included medical evacuations, ammunition and stores resupplies, forward air control,

leaflet dropping and the broadcasting of propaganda messages. Despite this multiplicity of roles, visual reconnaissance flights remained the unit's major commitment.

To be successful, these missions had to be flown at tree top height while the vulnerable and unarmed helicopters searched for signs of Viet Cong activity. Many such missions were uneventful, however, on occasion valuable intelligence information was collected. If located, the enemy invariably fired on the helicopter. To minimise this hazard as much as possible, the Sioux pilots would reduce height until they literally skimmed the tree tops in an attempt to reduce the number of guns which could be bought to bear on them.

During every major operation undertaken by Australian troops in Vietnam, one Sioux was allocated as the 'direct support' helicopter to each battalion. It was during one such mission for the

1st Battalion in the latter part of 1965 that 161 Flight suffered its first casualty when Captain B J Smith was wounded by small arms fire.

In May 1966, the unit relocated to Vung Tau in Phuoc Tuy Province where it was redesignated 161 (Independent) Reconnaissance Flight. From its new location, the unit was to support the activities of the 1st Australian Task Force. This move was only temporary however and, as it was desirable for the unit to be actually located with the Task Force, a second move to Nui Dat was effected in March 1967.

The 3200 square kilometres of Phuoc Tuy Province had to be searched regularly by 161 Flight and to facilitate this, the unit's establishment was increased to nine aircraft – six Sioux and three Cessna 180s. A Cessna Bird Dog – loaned from the US Army – was also operated.

On October 10 1966, Lieutenant W G Davies had a remarkable escape while conducting a road clearance mission over Route 15 when he was badly wounded by a sniper. His Sioux spun down out of control, struck power lines, crashed next to the road and somersaulted several times. Evacuated to an American hospital Lieutenant Davies underwent major surgery and, after being evacuated to Australia, made a full recovery.

In the heavily timbered terrain found in some of Phuoc Tuy Province, 161 Flight proved especially useful in evacuating wounded troops from clearings too small for larger helicopters. Sioux pilots were even known to land in minefields to evacuate badly wounded soldiers. Undoubtedly, the skill and heroism displayed during these extremely difficult operations saved many Australian lives.

In an attempt to reduce their helicopters' vulnerability, 161 Flight's ground staff devised a number of unofficial armament modifications for their Sioux. In one, twin M60 machine guns were fitted between the cross members of the helicopters skids. Another was effected by manufacturing a mount for a single M60 for use by the helicopter's observer. The fixed wing aircraft were similarly armed by substituting high explosive

rockets for the usual smoke type which were carried in underwing launchers to mark and attack targets.

One unusual aspect of 161 Flight's operations was the secondment of fixed wing pilots to the US Army's 54th Aviation Company at Long Thanh North. While on their six week attachment, the Australian pilots flew de Havilland Canada U-1A Otter single engined transports on freight missions over most of South Vietnam. Rotary wing pilots were not forgotten in this exchange programme and served with various American units where they flew helicopter types as diverse as the Hughes OH-6 Cayuse, Kiowa, Iroquois and Bell AH-1 Cobra.

While operating these machines several pilots were involved in major combats and one pilot, Second Lieutenant P Rogers, returned his helicopter in such a condition that, at the end of the day, it was written off. The ground staff were also, at times, attached to RAAF, USAF and US Army units where they operated both in their conventional role and as door gunners in the helicopters operated by these services.

In the latter part of 1969, Swiss built Pilatus Porters began to replace the Cessnas in Vietnam. Unfortunately, Porter operations began tragically, when during the first night training flight on December 3 1969, a Porter flown by Captain B C Donald and with Lieutenant A D Jellie on board, was shot down north of Nui Dat on the edge of the Binh Ba rubber plantation. Both occupants were killed. 161 Flight's only other fatality had occurred on May 23 1968 when Major G Constable was shot down in a Bird Dog.

Occasional night flare dropping missions became a feature of the unit's work once the Porters arrived and during one operation, on the night of August 11 1970, a Porter provided battlefield illumination for troops of the 8th Battalion. Nineteen enemy soldiers were killed and many others wounded in a successful ambush.

During 1971, modern Bell OH-58 Kiowa helicopters, leased from the US Government, began to replace the redoubtable Sioux that were used so effectively. By late 1971, 161 Flight was based at Vung Tau and covering the withdrawal of units

Sioux helicopters and a Porter on Nui Dat's Luscombe field.

from Nui Dat to Vung Tau whence they embarked for Australia.

161 Flight suffered Australia's last aerial loss of the Vietnam War when a Kiowa flown by Lieutenant G Steel (one of four New Zealand Army pilots to fly with the unit) was forced down by ground fire. The damaged helicopter was recovered and handed over to the US Army for repair but it is believed that the Kiowa was scrapped instead.

By Christmas 1971 most of 161 Flight had returned to Australia and only a small detachment of pilots, maintenance personnel and the Kiowas remained. On March 2 1972 they were returned to the Americans. A few days later the remaining personnel departed for Australia. 161 Reconnaissance Flight disbanded at Oakey later that year.

In Vietnam, 161 Flight attained a reputation for heroic flying and excellence in aircraft maintenance. Consequently, numerous decorations were awarded to individual personnel and, as a whole, the unit was much later (on July 4 1991) awarded the United States Army Meritorious unit Commendation for its service in support of the 173rd Airborne Brigade between 1965-1966.

During late 1973, 161 Reconnaissance Squadron reformed at Holsworthy with Australian built Kiowas. The unit provides tactical reconnaissance, artillery spotting, forward air control and medical evacuation support for the Australian Army. Additionally, the unit has been called upon to assist in a variety of civil aid tasks such as aeromedical evacuations, flood relief tasks and searches for lost civilians.

In December 1994, 161 Squadron moved to Darwin where it continues its operations. New helicopters, a number of which are being assessed as the Kiowa's replacement are due to enter service at the end of the 1990s.

Raised: *June 1965*
Aircraft: *Sioux, Cessna 180, Bird Dog, Porter, Kiowa*
Locations: *Amberley: June 1965 – Sept 1965*
Bien Hoa: September 1965 – May 1966
Vung Tau: May 1966 – March 1967
Nui Dat: March 1967 – August 1971
Vung Tau: August 1971 – March 1972
Oakey: March 1972
Disbanded:
Holsworthy: December 1973 – December 1994
Darwin: December 1994 – present

---

162 SQUADRON

162 (Independent) Reconnaissance Flight was raised at Amberley in September 1970.

Equipped with Sioux light observation helicopters the unit's role was to provide a tactical reconnaissance capability for combat elements of the Army. With its versatile helicopters, however, the unit has undertaken a variety of army support roles including casualty evacuation, artillery spotting and ammunition and stores resupply. Civil aid tasks such as search and rescue, medical evacuation and flood relief work have also been conducted on a regular basis.

In 1971, a detachment began operations from Townsville, to where the entire unit moved in October of that year. Retitled 162 Reconnaissance Squadron in late 1973, today the unit operates its Australian built Kiowa light observation helicopters in support of Townsville based units in north Queensland, and on occasion, in other areas.

Raised: *September 1970*
Aircraft: *Sioux, Kiowa*
Locations: *Amberley: Sept 1970 – Oct 1971*
Townsville: October 1971 – present

---

162 Squadron has operated Kiowas out of Townsville for many years in the reconnaissance role.

171 SQUADRON

171 (Air Cavalry) Flight was originally raised at Amberley on July 7 1968, under the command of Major D S Bell. Moving to Holsworthy early in the following year, the small Sioux and Cessna equipped unit supported the activities of the Army in NSW and Victoria before disbanding in 1973.

Equipped with Kiowa helicopters, 171 Command and Liaison Squadron was raised at Oakey the following year. Initially the unit flew in the reconnaissance and support roles implied by its title. Since re-equipping with former RAAF UH-1H Iroquois utility helicopters in 1990, 171 Squadron operates the full gambit of army support missions that the RAAF formerly undertook. These include troop insertion/extractions, casualty evacuation, reconnaissance and resupply of ammunition, food and equipment to troops in the field. In these roles, 171 Squadron regularly deploys from its home base, often operating from forward locations with few or no maintenance and domestic facilities.

As with the Army's other rotary wing units, 171 Squadron regularly supports the civil community during bushfires, floods and other emergencies. The Iroquois played a major role in the devastating Queensland floods around Charleville in 1990. Since then, 171 Squadron has conducted a number of successful mercy flights during which it has rescued lost or injured civilians. Humanitarian work has also taken 171 Squadron overseas when a detachment of helicopters and personnel were despatched to Fiji to assist that island nation which had been devastated by Cyclone Kina.

Redesignated an Operational Support Squadron in 1993, 171 Squadron continues to furnish valuable support both to the Defence Force and civil community.

Raised: *July 7 1968*
Aircraft: *Sioux, Cessna 180, Kiowa, Iroquois*
Locations: *Amberley: July 1968 – Jan 1969*
Holsworthy: January 1969 – 1973
Disbanded:
Oakey: 1974 – present

Equipped with floats this Army Iroquois alights on Burdekin Dam.

173 SQUADRON

173 (General Support) Squadron was raised at Oakey on February 17 1974. Commanded by Captain D B Coffey and equipped with Swiss built Porter aircraft, the new unit conducted a variety of tasks in support of the Army's Field Force. Four years later the unit also acquired Australian designed and built Nomad twin engined STOL transports.

173 Squadron participated in numerous exercises and its Porters, with their incredible short takeoff and landing ability, proved adept at a variety of tasks including tactical reconnaissance, casualty evacuation, troop transport, liaison tasks, artillery spotting, supply dropping, battlefield illumination and the delivery of supplies. Another important role undertaken was that of aerial survey photography in Irian Jaya, the Solomon Islands, Papua New Guinea and, more recently, in northern Australia. In 1990, a detachment of Porters participated in flood relief work

in the Charleville area where the small Army aircraft conducted search operations and dropped supplies to stranded farmers.

During November 1992 the Porters were withdrawn from service and in March 1993 the unit's title was changed to 173 (Surveillance) Squadron, to reflect its roles of surveillance, survey work and even paratrooping. Roles carried over from the Porter days include reconnaissance, liaison and transport tasks. Like the Porter before it, the Nomad possessed an excellent short field capability – a feature which assisted in its varied roles. The Nomads were grounded in late 1994 over concerns about their safety and suitability for military tasks. In mid 1995 the Army hoped to buy a new type, and in the interim had leased two Embraer Bandeirantes.

A 173 Squadron GAF Nomad.

Raised: *February 17 1974*
Aircraft: *Porter, Nomad, Bandeirante*
Locations: *Oakey: February 1974 – present*

5 AVIATION REGIMENT

Formed under the command of Lieutenant Colonel W J Mellor at Townsville on November 20 1987, 5 Aviation Regiment's role is to operate a large fleet of Sikorsky Black Hawk S-70A and Bell UH-1H Iroquois battlefield helicopters in support of Army activities.

Initially, the new Regiment took over those Black Hawks operated by the RAAF's 9 Squadron and due to an initial lack of specialist trained Army personnel, many RAAF aircrew and technical personnel – including virtually all of 9 Squadron – were integrated into the Regiment during the first few years of its existence.

5 Aviation Regiment was divided into two squadrons – A and B – both operating Black Hawks while a semi independent troop operated six Iroquois gunship attack helicopters. With the arrival of four Boeing CH-47D Chinook heavy lift helicopters in 1995, however, the gunships and Chinooks form the Regiment's C Squadron. The Chinooks, which are updated former RAAF CH-47Cs, substantially increase 5 Aviation Regiment's troop and artillery lift capability.

5 Aviation Regiment regularly participates in exercises providing troop lift, medical evacuation and stores resupply sup-

port for Army units. Another important role undertaken is to provide a battlefield airlift capacity for artillery units. The Black Hawks can deploy the 105mm howitzers, ammunition, equipment and personnel of artillery units over inaccessible areas particularly in northern Australia and, when their tasks are completed, return them either to their main bases, their transport vehicles or other forward positions.

5 Aviation Regiment has also been called upon

Army Black Hawks over their home town of Townsville.

January 1994 New South Wales bushfire disaster. The Black Hawks undertook a variety of tasks in this emergency including the transportation of firefighters and their equipment, evacuating trapped civilians and fire spotting. Most vital of all perhaps was the part played by the Regiment in fire bombing both on the main fire fronts and in inaccessible areas which could not be reached from the ground.

The Black Hawks have also participated in numerous rescue operations, one of the most significant occurring in April 1991 when 23 tourists were picked up from a capsized catamaran off the Queensland coast.

From a defence point of view, to date the Regiment's most important task was the deployment of six Black Hawks and 109 personnel from B Squadron to support United Nations peacekeeping operations in Cambodia.

These helicopters arrived in Cambodia in May 1993 and, liveried in UN white and fitted with armour plating and door mounted machine guns, supported the activities of the 22,000 United Nations peacekeeping troops in that country. Conditions were chaotic and, while deployed, a number of United Nations personnel were killed or wounded. Consequently, the Black Hawk crews had to use extreme caution when flying in most areas – the threat of ground fire being very real as was the risk of detonating land mines in some of the smaller landing zones.

The role of B Squadron was to resupply UN posts along the Thai – Laotian border, the transportation of electoral officials and ballot boxes for the UN sponsored national elections and the provision of a standby aeromedical evacuation helicopter. In the latter role many evacuations of UN soldiers and civilians were made. On at least three occasions, B Squadron helicopters were fired upon and in one of these a Black Hawk was slightly damaged by small arms fire. B Squadron returned to Townsville in July 1993.

With its ability to deploy infantry and artillery units over long distances, and operating one of the world's most advanced battlefield helicopters, 5 Aviation Regiment will continue to provide vital support for the Australian Defence Force into the future.

Formed: *November 20 1987*
Aircraft: *Black Hawk, Iroquois, Chinook*
Locations: *Townsville: Nov 1987 – present*

C Squadron, 5 Aviation Regiment, is custodian of the Army's Chinooks. (Helitech)

AIRFIELD LOCATIONS

Abeele – France
Abou Kir – Near Alexandria, Egypt
Abu Sueir – Delta Region of Egypt
Achofields – South of Sydney, NSW
Agnone – Eastern Sicily
Aitkenvale – Near Townsville, Queensland
Alem el Gzina – On Gulf of Sirte, Cyrenaica
Alghero – North West Sardinia
Almaza – Near Cairo, Egypt
Amberley – Inland from Brisbane, Queensland
Amel-el-Chec – Libya, 80km from Marble Arch
Amman – North east of Dead Sea, Jordan
Amriya – Inland from Alexandria, Egypt
Andir – Western Java
Andreas – Isle of Man, United Kingdom
Antelat – Inland from Beda Fomm, Cyrenaica
Aqir – East of Haifa, Palestine
Archerfield – Brisbane, Queensland
Auchel – Near Lille, France
Baginton – Southern England
Baileul – North west of Lille, France
Balikpapan – East coast of Borneo
Bankstown – Sydney, NSW
Bari – North east coast Italy
Bassingbourn – South east England
Batavia – North east Java
Batchelor – South of Darwin, NT
Beauvais – Normandy region, France
Bien Hoa – North of Saigon, South Vietnam
Belaudah – On Gulf of Sirte, Libya
Benina – Inland from Benghazi, Cyrenaica
Berca – Near Derna, Cyrenaica
Biak – Inland Dutch New Guinea
Binbrook – North east England
Bofu – Southern Honshu, Japan
Bohle River – Near Townsville, Queensland
Bone – Algerian Coast near Tunisian border
Bottesford – North of London, England
Bouvincourt – West of St Quentin, France
Bowen – North Queensland, South of Cairns
Breighton – North of London, England
Broome – North west Western Australia
Bundaberg – North of Brisbane, Queensland
Burg-el-Arab – Near Alexandria, Egypt
Butterworth – Near Penang, Malaysia
Cairns – North Queensland coast
Camden – Near Sydney, NSW
Camp Bearcat – Bien Hoa Province, Vietnam
Camp Blackhorse – Long Kanh Province, Vietnam
Cape Gloucester – Western tip of New Britain
Castel Benito – Near Tripoli, Libya
Castlereagh – Near Sydney, NSW
Cecil Plains – Inland southern Queensland

Cervia – North Coast of Italy
Changi – Singapore Island
Charleroi – East of Brussels, Belgium
Cirencester – Gloucester, England
Coffs Harbour – Northern coast of NSW
Coltishall – South east England, north of Ipswitch
Cooktown – Coast of north Queensland
Coomalee Creek – South of Darwin, NT
Corrunna Downs – North west Western Australia
Culdrose – Cornwall, England
Cunderin – Inland southern Western Australia
Cutella – Italy, north of Foggia
Cuers – France
Daly Waters – South of Darwin, NT
Dallachy – North East Scotland
Damascus – Syria
Darwin – Northern Territory
Deir-el-Belah – Palestine, north west of Gaza
Deniliquin – Southern New South Wales
Detling – North east of Maidstone, England
Deurne – France
Dovai – South east of Lille, France
Dong Tam – Dinh Tuong Province, South Vietnam
Drem – Near Edinburgh, Scotland
Driffield – North east England near North Sea
East Sale – Eastern Victoria
East Kirkby – North east England
Edinburgh – Near Adelaide, South Australia
Eglinton – Northern Ireland
El Bassa – Egypt
El Hamma – Tunis, Tunisia
El Daba – Egypt
El Assa – Libya near Tunisian border
El Adam – Inland from Tobruk, Cyrenaica
El Mejdel – North of Gaza, Palestine
Essendon – Melbourne, Victoria
Fairbairn – Canberra, ACT
Falerium – North of Rome, Italy
Fano – North east coast of Italy
Fassberg – Germany
Fayid – Egypt on the Suez Canal
Feltwell – North of London, England
Fenton – Near Adelaide River, NT
Finschafen – Tip of Huon Gulf, New Guinea
Foggia Main – Inland Italy south east of Cassino
Ford – West of Littlehampton, south coast of UK
Foulsham – East England near North Sea
Fouquorolles – France
Gambut – Between Tobruk and Bardia, Cyrenaica
Gamil – Near Port Said, Egypt
Garbutt – Townsville Queensland, Australia
Gardabia Mam – Libya, east of Tripoli
Gatow – Berlin, Germany

Gawler – Near Adelaide, South Australia
Gazala – Between Derna and Tobruk, Cyrenaica
Gerawla – Near Mersa Matrah, Egypt
Gibraltar – Southern tip of Spain
Goodenough Is – Southern tip of New Guinea
Got-es-Sultan – Inland from Benghazi, Cyrenaica
Gould – South of Darwin, Northern Territory
Gove – West coast Gulf of Carpentaria, NT
Grottaglie – Near Tarano, Italy
Guidonia – Near Rome, Italy
Guildford – Near Perth, Western Australia
Halfar – Malta
Hamraiet – Inland from the Gulf of Sirte, Libya
Harlaxton – Near Gratham, England
Hawkinge – East coast of England near Dover
Hellemmes – East of Lille, France
Helwan – Near Cairo, Egypt
Higgins Field – Tip of Cape York Peninsula
Holsworthy – Near Sydney, NSW
Hollandia – North east coast of Dutch New Guinea
Holme-on-Spalding-Moore – Hornsea, England
Horn Island – Above the tip of Cape York, Qld
Hornchurch – Near London, England
Hughes – South of Darwin, Northern Territory
Ibsley – Southern England near Winchester
Idku – Near Alexandria, Egypt
Iesi – North coast of Italy
Ipoh – Western central Malaya
Iwakuni – Southern Honshu, Japan
Jondaryan – Inland southern Queensland
Julis – Southern Palestine
Jurby – Isle of Man, United Kingdom
Kairouan – Tunisia
Kalidjati – Western Java
Kantara – Egypt on Suez Canal
Karumba – Western side of Cape York Peninsula
Kasarfeet – Egypt on Suez Canal
Kenley – England, south of London
Kimpo – South Korea near Seoul
Kingaroy – Inland southern Queensland
Kiriwina – Trobriand Islands
Kirton-in-Lindsay – Eastern England
Kota Baharu – North east Malaysia
Kuala Lumpur – South west Malaya
Kunsan – West coast of South Korea
La Bellevue – South west of Arras, France
Labuan – Island off north west coast of Borneo
Lae – East coast of New Guinea
Laha – Ambon Island, Netherlands East Indies
Langham – East coast of England
Lasham – East coast of England
Lavariano – Italy
Laverton – Melbourne, Victoria
Leconsfield – North east England near Hull
Leighterton – Gloucester, England
Leuchars – East coast of Scotland
Leyburn – Inland southern Queensland, Australia
LG "Y" – Egypt

LG 75 – Inland from Sidi Barrani, Egypt
LG 76 – Inland from Sidi Barrani, Egypt
LG 79 – Near Bardia, Cyrenaica
LG 91 – Egypt
LG 101 – Inland from Mersa Matruh, Egypt
LG 106 – Egypt
LG 110 – Inland from Sidi Barrani, Egypt
LG 132 – Near Maddalena, Egypt
LG 142 – Egypt
LG 167 – Egypt
LG 175 – Egypt
LG 222 – Egypt
LG 224 – Egypt
LG 237 – Egypt
Liettres – France
Serny – Near Lille, France
Livingstone – South of Darwin, NT
Long – Midway between Darwin & Katherine, NT
Longues – Normandy, France near "Gold" Beach
Lowood – Inland southern Queensland
Luqa – Malta
Lydda – Palestine, inland from Tel Aviv
Lympne – South east England near Dover
Macrossan – Inland north Queensland,
Madang – North east New Guinea
Maddalena – Cyrenaica near Egyptian border
Mallala – North west of Adelaide, South Australia
Manbulloo – South west of Katherine, NT
Manston – Near Dover, south east England
Maraua – Near Barce, Cyrenaica
Marble Arch – On Gulf of Sirte, Libya
Mareeba – Inland from Cairns, north Queensland
Martlesham – South east England near Ipswitch
Martuba – Near Derna, Cyrenaica
Marylands – Western Australia
Matlaske – Near Swannington, south east England
Medenine Main – Near Tatalin, Tunisia
Melville Island – north east of Darwin, NT
Menangle – Queensland
Merauke – South coast of Dutch New Guinea
Mersa Matruh – Egyptian coast near Cyrenaica
Methringham – South of Hull, England
Methwold – Inland from Dunwich, England
Mileni – North of Foggia, Italy
Milne Bay – Southern tip of New Guinea
Minchinhampton – Gloucester, England
Molesworth – North of London, England
Momote – Los Negros Island
Montigny – East of Douai, France
Moratai – Island to the south of the Philippines
Mount Batten – Southern UK on English Channel
Nadzab – Northern New Guinea on coast
Nadzab – Mouth of Markham River, New Guinea
Namlea – Buru Island, Netherlands East Indies
Narromine – Central New South Wales
Nefatia – Tunisia
Noemfoor – Island north of Dutch New Guinea
Nowra – South coast of New South Wales

Nui Dat – Phuoc Tuy Province, South Vietnam
Oakey – Inland southern Queensland, Australia
Oban – North west Scotland
Onslow – Near Exmouth Gulf, Western Australia
Pachino – Southern Sicily
Palembang – Inland southern Java
Parafield – North of Adelaide, South Australia
Pearce – Near Perth, Western Australia
Pembroke Dock – Western England on Irish Sea
Perranporth – Near south west tip of England
Phan Rang – Near Cam Ranh Bay, Sth Vietnam
Piva North – Torikina, Bougainville Island
Pohang – East coast of South Korea
Point Cook – Near Melbourne, Victoria
Pont-a-Marq – South of Lille, France
Poole – On English Channel, England
Poretta – Corsica
Port Moresby – Southern New Guinea
Potshot – Near Exmouth Gulf, Western Australia
Poulainville – North of Amiens, France
Predannack – Southern tip of England
Premont – North east France
Protville – Tunisia near Bizerta
Proyart – Near Albert, France
Pusan – South east South Korea
Qasaba – Near Mersa Matruh, Egypt
Qassasin – Delta region of Egypt
Rafa – South west of Gaza, Palestine
Ramleh – South of Tel Aviv, Palestine
Rathmines – Coast of New South Wales
Rayak – Central Lebanon
Redhill – South of London, England
Richmond – Near Sydney, New South Wales
Rochford – North of Southend, England
Rosh Pina – Near Lake Tiberias, Palestine
Ross River – Near Townsville, Queensland
San Angelo – South of Cassino, Italy
Sattler – South of Darwin, Northern Territory
Savy – West of Arras, France
Sedada – Libya
Selatar – Singapore Island
Sembawang – Singapore Island
Semplak – Western Java
Serny – Near Lille, France
Serragia – Corsica
Shallufa – Near Suez Canal, Egypt
Shawbury – England
Sidi Haneish – Near Mersa Matruh, Egypt

Sidi Azeiz – Inland from Bardia, Cyrenaica
Sidi Mahmound – Near Tobruk, Cyrenaica
Skaebrae – Northern tip of Scapa Flow, England
Skellingthorpe – South of Hull, England
Solluch, Near Benghazi, Libya
South Carlton – England
St Merryn – United Kingdom
St Georges Basin – Near Jervis Bay, NSW
St Catherine – Corsica
St Jean – Near Acre on the coast of Palestine
Strathpine – Near Brisbane, Queensland
Strauss – South of Darwin, Northern Territory
Sugai Pattani – West coast of Malaysia
Swannington – Eastern England near North Sea
Swinderby – North of London, England
Tadji – Northern New Guinea on coast near Aitape
Taegu – Central southern South Korea
Takali – Malta
Tarakan – Small island off Borneo
Tengah – Singapore Island
Tera – Central Italy
Tern Hill – England
Tetbury – Gloucester, England
Thornaby – East coast of England
Tindal – South of Darwin, Northern Territory
Tmimi – Near Derna, Cyrenaica
Tocumwal – Inland NSW on Victorian border
Toogoolawah – North west of Brisbane, Qld
Toowoomba – Inland from Brisbane, Qld
Townsville – North Queensland, Australia
Ubon – Thailand near border with North Vietnam
Valley – West coast of Wales
Villers-Bocage – Normandy, France
Vung Tau – Phuoc Tuy Province, South Vietnam
Waddington – North east England
Wadi Natrum – South of Alexandria, Egypt
Wagga – Inland southern New South Wales
Warloy – Near Baizieux, France
Wehi Sheikh Nuran – South of Gaza, Palestine
Wigsley – South of Poole, England
Williamtown – North of Newcastle, NSW
Wooloomanata – Near Geelong, Victoria
Wunsdorf – Berlin, Germany
Yatesbury – East of London, England
Yeovilton – United Kingdom
Yonpo – East coast of North Korea
Zuara – Libya

THE *EXCITING* STORY OF AUSTRALIA'S AIR WAR IN VIETNAM

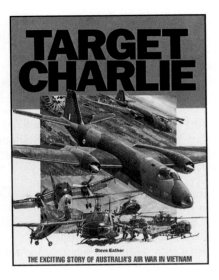

Over the period of a decade some 57,000 Australian and New Zealand service personnel served in Vietnam, yet the story of their eight year air war has rarely been told.

Target Charlie is about that air war, the people in it and their achievements, challenges and disappointments in an unforgiving and hostile land. A thoroughly readable text is supported by a wide range of photos, maps, statistical data and tables not previously published in a single book.

Target Charlie is not only an excellent ready reference work on Australia's air war in Vietnam but also a vital account of a period in Australian history that should not be forgotten.

Target Charlie, for the first time ever, includes a comprehensive analysis of Australia's air war in Vietnam backed up by individual chapters on the operations of 35 Sqdn (DHC Caribou), 161 Independent Reconnaissance Flight (Sioux, Kiowa, Cessna 180, Pilatus Porter), 9 Sqdn (Bell Iroquois), 2 Sqdn (Canberra), Royal Australian Navy Helicopter Flight – Vietnam, plus Australia's involvement with various US Forward Air Control units and fast jet squadrons flying the F-4 Phantom. Also included are maps detailing the area of operations, a chronology of events table, a background to the historical causes of the Vietnam War, listings of Australian aircraft lost or damaged during the conflict plus a roll of honour of those who paid the ultimate price.

Target Charlie is truly the story of Australia's air war in Vietnam and is an essential reference work for anyone seriously interested in Australia's military heritage.

✄ -

YOU'VE READ THE SQUADRON HISTORIES, NOW READ ABOUT THE INDIVIDUAL AIRCRAFT THAT MADE THESE SQUADRONS WHAT THEY WERE!

CELEBRATING THE AIRCRAFT THAT MADE AUSTRALIAN MILITARY HISTORY

EACH BOOK IS ESSENTIALLY THREE BOOKS IN ONE!

★ Individually narrates the exciting development history of each particular aircraft including their entry into general service worldwide.

★ Documents in detail each aircraft's adaptation to Australian service and describes their war record, where applicable, in full.

★ Unique tables document each aircraft individually, listing its A series registration, individual version or batch number, where applicable, date of delivery and, most importantly, its eventual fate. This never before available data will provide a unique source of reference material for years to come.

★ Over 70,000 words of text backed up by hundreds of photos in each book, most of which have never before been published. These excellent books are worth the modest purchase price for the pictorial content alone.

--✂-----------

NOT TO BE
M I S S E D !

▷ *an excellent year round gift if ever there was one*

To commemorate the completion of our **Australian Airpower** series we have produced a special high quality display poster depicting the cover artwork of the 40 principal aircraft that have faithfully served Australia through both peace and war.
As a special bonus this 60cm x 84cm full colour poster comes with a free eight page booklet featuring the career details of each type. This free booklet includes data on years in service, quantity in service, various career aspects and other relevant data that will enable you to easily and enjoyably appreciate the exploits of each of the 40 aircraft that have been highlighted in our popular **Australian Airpower** series.

❝ THE *IDEAL* ADDITION TO YOUR STUDY, PLACE OF WORK OR A SPECIAL HIGHLY VALUED GIFT ❞

CELEBRATING THE AIRCRAFT THAT HAVE PROUDLY SERVED AUSTRALIA THROUGH BOTH PEACE AND WAR FOR MORE THAN HALF A CENTURY

ONLY $19 POST FREE
(Domestic sales only, add $4 for overseas seamail orders)

✂

WINGS ACROSS THE SEA

Ross Gillett

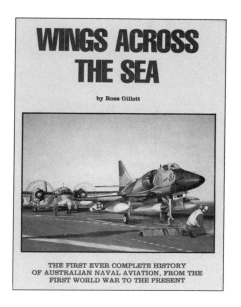

This 164 page book features the complete history of three quarters of a century of Australian naval aviation, through both peace and war. An excellent and informative narrative in addition to being of immense reference value. It has complete tabulated data on every aircraft to have served with the RAN including individual squadron histories and a chronology of major RAN FAA events.

Wings Across the Sea is illustrated with more than 290 photos including 34 in glorious colour, the vast majority of which have never before been published.